Troubleshooting & Repairing
Audio & Video Cassette
Players & Recorders

I dedicate this book to the three Davids, David Held, David Mark, and David Davidson, who have helped on several electronic books the past 22 years.

Troubleshooting & Repairing Audio & Video Cassette Players & Recorders

Homer L. Davidson

TAB Books
Division of McGraw-Hill, Inc.
New York   San Francisco   Washington, D.C.   Auckland   Bogotá
Caracas   Lisbon   London   Madrid   Mexico City   Milan
Montreal   New Delhi   San Juan   Singapore
Sydney   Tokyo   Toronto

5359708

pbk 5 6 7 8 9 10 11 12 13 14 DOH/DOH 9 9 8 7 6 5 4
hc   1 2 3 4 5   6   7   8   9 10 DOH/DOH 9 9 8 7 6 5 4 3 2

Library of Congress Cataloging-in-Publication Data

Davidson, Homer L.
    Troubleshooting & repairing audio & video cassette players &
recorders / by Homer L. Davidson.
        p.   cm.
    Includes index.
    ISBN 0-8306-4259-5 (hard)   ISBN 0-8306-4258-7 (pbk.)
    1. Magnetic recorders and recording—Cassette recorders-
-Repairing. 2. Video tape recorders and recording—Repairing.
I. Title. II. Title: Troubleshooting and repairing audio and video
cassette players & recorders.
TK7881.6.D39   1992
621.388′337—dc20   92-9559
                                                    CIP

Acquisitions Editor: Roland S. Phelps
Book Editor: Andrew Yoder
Director of Production: Katherine G. Brown
Book Design: Jaclyn J. Boone
Cover Photograph: Thompson Photography, Baltimore, Md.   TAB1
Cover Design: Holberg Design, York, Pa.   3795

Contents

4 Servicing Boom-box Cassette Players 85

5 Troubleshooting Portable AM/FM Cassette/CD Players 107

10 Maintaining Double Cassette Decks   **265**

11 VCR Repairs You Can Make   **287**

12 Troubleshooting Stereo Cassette Decks   **313**

Acknowledgments

A great deal of thanks for all the help of the electronic technicians and cassette manufacturers, whose cassette players are mentioned throughout the book. Also, a special thanks to Radio Shack for service data and schematics found in several chapters. Their assistance helped make this book possible.

Introduction

*T*he cassette player is used in every room of the home or office, outdoors, in the auto, and while running for more exercise. The purpose of this book is to help the homeowner, tinkerer, hobbyist, beginner, and electronics student to learn how each player operates and how to make simple repairs. Even you, the cassette owner, can place that unit back into service.

This book has 15 chapters on cassette players, starting with simple player basics, personal tag-along, boom box, cassette/CD, mini-microcassette recorders, professional recorders, compact cassette players, and the cassette players that are found in the combination AM/FM phonograph.

The first chapter shows how to make transistor and IC tests with only one test instrument, In fact, with a multipurpose digital meter (DMM), you can make most of the different sound repairs. Then, add a couple of build-your-own test-equipment projects and you're in business, so to speak.

How to repair those tag-along, personal and portable cassette players is given in chapters 2 and 3. The boom-box repairs and hints are found in chapter 4. Problems with the microcassette and professional recorders are given in chapters 6 and 7.

The overworked cassette player in the car finally breaks down and needs repair. Chapters 8 and 9 provide troubleshooting tips on how to service the car stereo cassette and CD players. The auto cassette/CD timer combination might consist of a cassette/tuner, cassette/receiver, cassette/CD controls, under-dash CD player, or changers in the trunk.

Separate recording and playback features, high-speed dubbing and soft-door release are found in chapter 10 with double cassette deck maintenance. Simple VCR cassette repairs are given in chapter 11. Troubleshooting the stereo cassette decks and servicing the compact-rack cassette players are found in chapters 12 and 13.

Just about anyone can make the most simple camcorder cassette repairs with chapter 14. Knowing the different camcorder formats, cassette problems, and clean up methods can solve a lot of camcorder problems. Locating that special part or component is easy with the correct manufacturer's address, listed in chapter 15.

Although certain cassette problems are provided in a given chapter, the same problem might be found in another chapter and related to the cassette player you are now servicing. Cassette troubleshooting and servicing methods are all here, so you can bring that cassette player back to life. Just have some fun and begin with the basics in chapter 1.

Chapter **1**

Cassette player basics

*T*he cassette player is available in several different general shapes and sizes with different features. You can find the cassette player in every room of the house, in the auto, and at the beach (FIG. 1-1). Thousands of them are out there, broken down, collecting dust, in need of repair. Troubleshooting and repairing your own cassette player can be quite rewarding. Besides, it's lots of fun and you can save a few bucks in the process.

Servicing cassette players can be done by anyone who is handy with tools and can read a schematic diagram (FIG. 1-2). The electronic circuits are broken down so that everyone can understand them. If you have a little knowledge of electronics, if you are a novice or if you are an electronics student, you can make most repairs found in this book. Very few tools and test instruments are needed. In fact, only one digital-multimeter (DMM) with many functions can do the job.

REQUIRED TEST INSTRUMENTS

Several screwdrivers, a pair of long-nose pliers, and side cutters do the bulk of the work. A set of jeweler's screwdrivers take care of those tiny screws and bolts (FIG. 1-3). The small magnifying glass helps to locate small parts and soldered terminals. If you do not have a VOM or a DMM, purchase one of the new digital multimeters that can check voltage, resistance, current, capacitance, and diodes with a frequency counter.

Today, several types of pocket DMMs can do all the tests needed to repair that broken cassette player. The tester might cost up to $150.00. The BK test-bench DMM (FIG. 1-4), model 388-HD, can check diodes, transistors, logic, frequency, current, capacitance, resistance, and voltage. Read carefully how each test is made from the manufacturer's literature included with the DMM.

This DMM can check transistors and diodes with the diode test. npn and pnp transistors can be tested by plugging into a small transistor socket. The frequency counter has three different frequency ranges: 2 kΩ, 20 kΩ, and 200 kΩ. Current

1-1 The small portable and boom-box cassette players are the most popular units around the house.

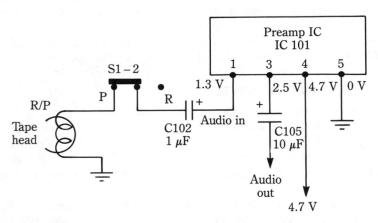

1-2 If you can read the simple schematic diagrams and handle a few tools, you're in business.

from 200 μA to 20 A can be checked on five different ranges (FIG. 1-5). Small capacitors can be tested from 2 nF to 20 μF. Seven resistance ranges vary from 200 Ω to 2000 MΩ. dc voltages vary from 200 mV up to 1000 V, and the ac voltage range goes up to 750 V.

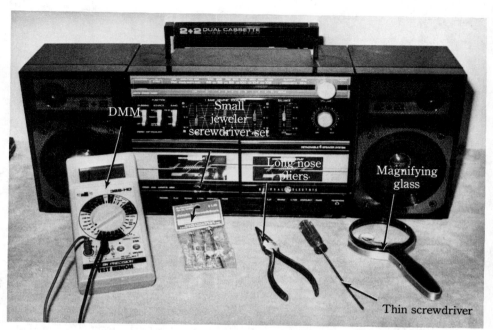

1-3 A few small tools might be required to put that cassette player back in tip-top shape.

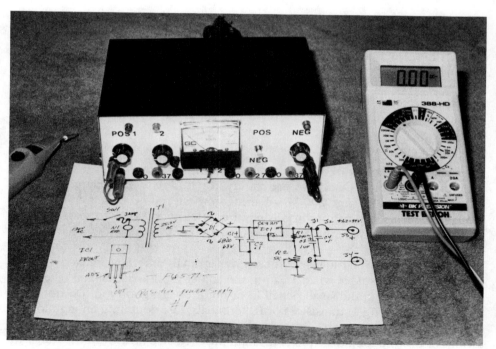

1-4 The B & K test bench DMM can make all the tests given in these pages.

1-5 Here, the DMM is used in series with batteries to check the total current (16.65 mA) of a pocket cassette player.

Besides taking critical voltage measurements upon transistors and IC components, transistors and diodes can be tested with the diode-junction tests or transistor-gain measurements. The low ac voltage range can be used in azimuth head alignment. Speed problems can be checked with the frequency-counter test. Total current drain of the small battery-operated cassette player can indicate leaky components. Defective or unknown small capacitors can be checked with this small DMM.

SOLDER EQUIPMENT

The small 30-W soldering iron is needed to remove transistor and small board components (FIG. 1-6). Of course, the battery iron is ideal for surface-mounted and IC terminals, and the large 200-W soldering iron can remove larger components, mesh and shields, but neither iron is required. If you are going to service a large group of cassette players, the controlled-temperature iron is handy to have on the service bench (FIG. 1-7).

A small pencil iron can solder those tiny terminals or remove melted soldered connections with solder wick. Heating the solder-mesh material can take a few seconds longer with the small pencil iron, but it does a good job. Do not apply the iron tip too long to transistors or IC terminals; just do it long enough to melt the solder and make a good soldered joint. Too much heat can damage the transistor or IC. Use the long-nose pliers to drain off excessive heat from the transistor leads.

1-6 The small 30-W soldering iron can make all soldered connections inside the cassette player.

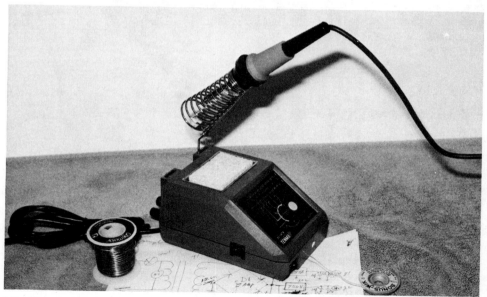

1-7 The temperature-controlled soldering iron is nice if you have a lot of soldering to do when repairing cassette players.

After locating a defective transistor or IC, lift the solder from each terminal with solder wick. Do not apply too much heat if the transistor is to be tested out of the circuit. Sometimes transistors can test bad in the circuit and test good when removed. Be careful not to apply too much heat to the small PC wiring, so as not to "pop" off the wiring from the board. The defective IC can be quickly removed by applying heat to a row of terminals with solder wick. Remove excess solder from the board with solder mesh after the part has been removed.

BATTERY PROBLEMS

When the portable cassette player appears to be dead or if the tape will not rotate, test each battery. These small batteries can be checked with a battery tester or with the voltage test of a VOM or DMM. Do not remove the batteries when taking voltage tests. Place the DMM test leads across each battery with the cassette switch turned on and a tape loaded. The battery can test close to normal when removed from the cassette player and with no load. The audio will become weak and the play speed will be slow when a 1.5-V battery drops to 1.25 V (FIG. 1-8).

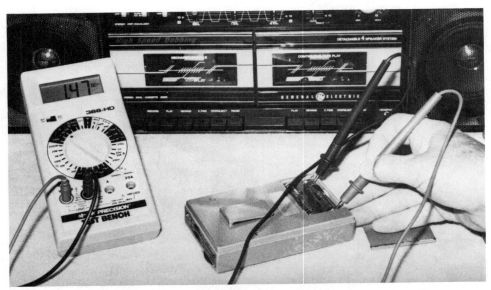

1-8   Check the batteries in the cassette player while operating to determine if they are weak and used up.

Replace the 9-V battery when the voltage drops below 7.5 V under load. You can locate only one dead or low-voltage battery with several other batteries. If all batteries were installed at the same time, remove and replace all of them when one drops below the required operating voltage.

Wipe the battery terminals on a cloth to clean off the contacts. Clean the battery contacts with cloth and alcohol while the batteries are removed. Inspect the battery terminals for broken or corroded terminals. Do not leave batteries within the cassette player if the unit is not used in a three-month period.

Rechargeable batteries should be charged when the speed slows down or if weak audio is noticed. Like the battery shaver, charge them up before using. Some people say to discharge nicad batteries before charging. But, this is not necessary, just charge them. When these batteries will not hold a charge for only a few minutes, discard the chargeable batteries. The defective rechargeable battery will not hold a charge very long within the cassette player. Overloaded circuits within the cassette player can cause the batteries to wear out in a short time.

TEST TAPES AND TENSION GAUGES

One or two test tapes are handy when checking audio stages, tape speeds, and head alignments. Tension gauges can be used to check the pressure roller, takeup torque, and tape tension (TABLE 1-1). Although these tension gauges are not necessary, they can speed up repairs. Today, some of these test tapes and tension gauges can be difficult to find. Try local electronic stores and cassette player manufacturer depots to locate them.

Table 1-1. Test Tapes and Frequencies

Frequency	Test cassette	Function
10 kHz	VTT – 658 MTT – 114 MTT – 216	R/P head azimuth
6.3 kHz	Standard	Head azimuth and sensitivity
3 kHz	MTT – 111	Tape speed adjust
1 kHz	MTT-118	Tape speed adjust
400 Hz	MTT – 150	Playback-level sensitivity

You can make your own 1- to 10-kHz test cassettes by applying the audio signal into the external microphone jack or by clipping the signal across the tape head terminals. Inject the signal from an audio signal generator or from a home-constructed sine-/square-wave signal oscillator given in home-built test equipment (FIG. 1-9). Check the exact frequency with the frequency counter of the DMM. Record at least one hour of audio signal on a new cassette for each test tape. Remove the input signal and play the recorded 1-kHz/10-kHz signal.

Always keep the volume low so that the audio input signal does not distort. Try another recording if too much volume, too little volume, or distortion is found on the cassette. These homemade test cassettes can be used for audio troubleshooting, azimuth head aligning, tape speed checking, and for locating weak audio stages.

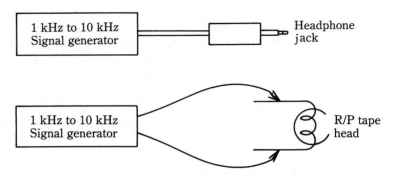

1-9   How to inject a 1-kHz signal into the cassette player to record a test cassette.

TROUBLESHOOTING WITH VOLTAGE AND RESISTANCE MEASUREMENTS

The defective transistor can be located with voltage and transistor measurements. Voltage measurements on each collector, base, and emitter terminal with common probe to ground can indicate a defective transistor. The open transistor can have a higher-than-normal collector voltage and no voltage on the emitter terminal (FIG. 1-10). An open emitter resistor or terminal can have zero measurement. Be careful not to short any two terminals together with the test probes.

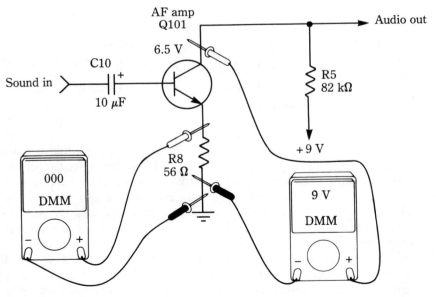

1-10   Higher collector and zero emitter voltage indicates that a transistor is open.

The leaky or shorted transistor can have close voltages on all terminals. Most transistors become leaky from emitter to collector (FIG. 1-11). Check the transistor with in-circuit transistor tests and then remove them from the circuit and take another leakage test.

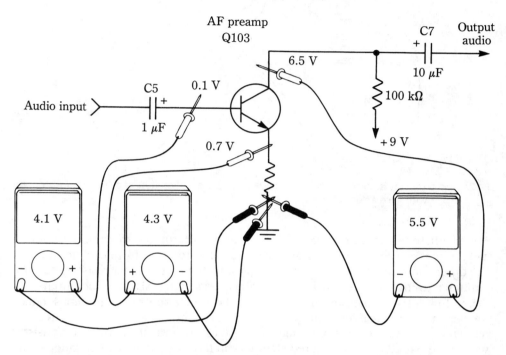

1-11 The critical voltages indicate that a transistor is leaky if the voltages on all of the terminals are quite close.

Another method to determine if the transistor is normal is to measure the bias voltage from the base to emitter terminals. Usually, the transistor is normal with either a 0.6- or 0.3-V measurement (FIG. 1-12). The npn and pnp silicon transistor has a 0.6 bias voltage and the pnp germanium transistor has a 0.3 bias voltage. The difference in both voltage measurements from base to ground and emitter to ground should equal the bias voltage of a normal transistor.

The intermittent transistor can show different identical voltage measurements on collector and emitter terminals, The voltage can quickly change with the intermittent transistor. Sometimes, when the transistor is in the intermittent state, when touched with a test probe, the voltage will return to normal, which shocks the transistor. The intermittent transistor might test open or leaky in the circuit and tests normal when removed from the PC board. If this is the case, replace the suspected transistor.

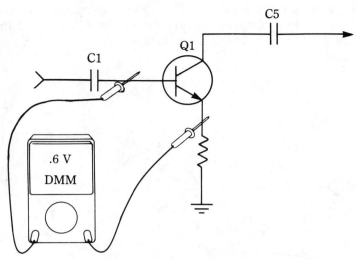

1-12 The transistor is normal with 0.6 V between the base and emitter of a silicon transistor and 0.3 V with germanium transistor.

Accurate resistance measurements of transistor terminals to ground can locate a defective transistor. Take critical resistance measurements from each terminal to the other can indicate a leaky transistor (FIG. 1-13). Open emitter resistors or poor emitter terminal connection results in no or real high-resistance measurements. When low resistant measurements are found on transistor or IC terminals, check the schematic for real low-ohm resistors, diodes, and coils within that circuit. A resistance measurement from each element to ground of an output transistor within the directly coupled stereo channel can indicate a defective channel. Compare the same resistance measurements with the good channel.

TRANSISTOR TESTS

The suspected transistor can be checked with a transistor tester or with the diode test of the DMM. The transistor tester can check the transistor in or out of the circuit. The transistor must be removed and plugged into the transistor socket with the B & K 388-HD DMM.

1. Set the Function/Range switch to the desired HFE (dc transistor gain) range (*pnp* for pnp transistors and *npn* for npn transistors).

2. Plug the transistor directly into the HFE socket. The sockets are labeled *E* (emitter), *B* (base), and *C* (collector).

3. Read the transistor HFE (dc gain) directly from the display.

Notice this B & K DMM will have an overrange symbol (1) when turned to diode test, logic, and resistance without probe connections in the circuit. All other ranges, such as resistance, frequency, current, capacitance, and voltage will

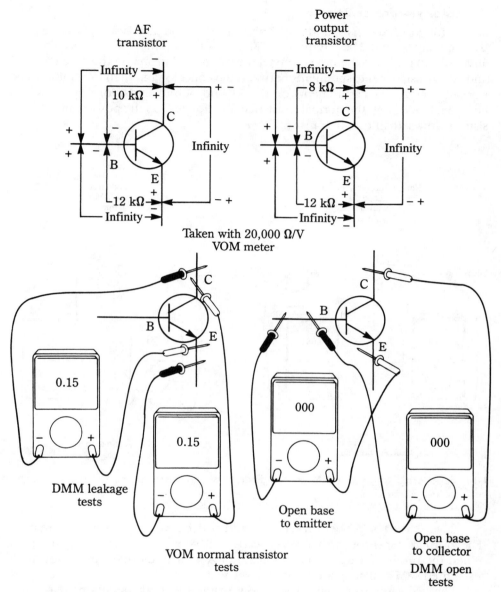

1-13 The resistance and leakage test of the AF and power-output transistors found in the cassette player with a 20,000 V/Ω VOM.

have a (000) indication without probe connections. The overrange symbol will show up if the range is over that in which the selector knob is turned to. The low ohm scale (200) can have a 00.1 resistance measurement on display with probes touching, which indicates resistance of the test leads to the meter circuit.

Transistor junction-diode tests

Set the function switch to diode symbol tests. With npn transistors, place the positive (red) probe to the base terminals. While testing pnp transistors, place the negative (black) probe to the base terminal. Leave the red probe on the base terminal and take resistance measurements between the collector and then the emitter. The normal resistance measurements on the AF and output transistors are shown in FIG. 1-14. Notice that the normal resistance junction test on the power output transistor is lower than the AF transistor.

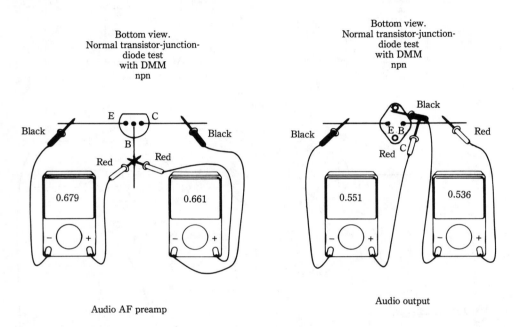

I-14 Take normal transistor DMM junction-diode tests with the red probe at the base terminal of an npn transistor.

Reverse the test probes on each test and notice if you receive a measurement. The normal transistor with reverse leads will show an overrange symbol (1). When the probes are accidentally touched together in diode or transistor-diode tests, the DMM available continuity buzzer will sound.

The leaky transistor will have a low reading in both directions (FIG. 1-15). Most transistors will short between the collector and emitter. The shorted transistor has only a fraction of an ohm short (0.015), but a higher leakage can show a reading in both directions. The transistor can become leaky between any two elements or all three. The leaky or shorted transistor will have a resistance measurement in both directions.

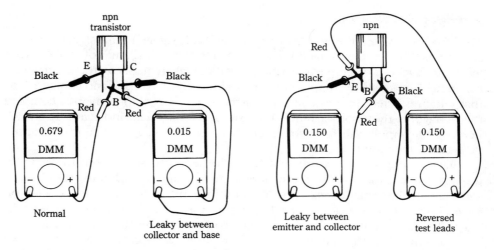

1-15 The leaky or shorted transistor will have a very low resistance in both directions.

You might find a transistor with a normal measurement between the base and emitter terminals and a high reading between the base and collector (FIG. 1-16). Replace the transistor with a high-resistance measurement, which indicates a high-resistance junction. Remember, both normal measurements between the base and collector, and the base and emitter, should only be a few Ω apart.

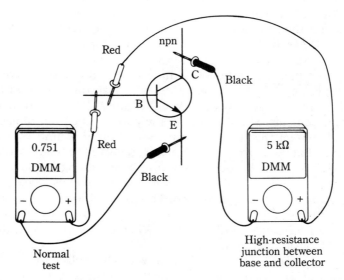

1-16 Replace the transistor if you find a high-resistance junction measurement is between the base and emitter or collector terminals.

IC TESTS

The suspected IC component can be located with signal-in and signal-out measurements. If the audio signal is traced to the input terminal and not at the output terminal, suspect a defective IC. Critical voltage measurements on the IC terminals can indicate a defective component. Measure the supply voltage and compare it to that which is listed on the schematic. If the schematic is not available, compare the voltage to the good channel. Low supply voltage can indicate a leaky IC. Remove the supply pin from the circuit with solder wick. Notice if the supply voltage has increased.

Take a resistance measurement from the removed supply pin to common ground. Replace the leaky IC if the measurement is under 1 kΩ. If the supply voltage remains the same, check each IC pin to ground with ohmmeter. When you have a low ohmmeter reading, check that same pin on the schematic to determine if a coil or low-ohm resistor is in the same path. If not, replace the leaky IC.

Sometimes, with ohmmeter tests, you can locate a change in a resistor or leaky capacitor from pin to common ground. Make sure the IC is defective by taking signal-in and signal-out tests, with critical voltage and resistance measurements. Always compare these measurements with the good channel in the stereo amplifier. Replacing the IC requires a little more time than replacing a transistor.

Transistor and IC replacement

After locating the defective transistor in the amplifier, the transistor must be removed and replaced. Most transistors found in audio circuits can be replaced with universal replacement transistors if the original part is not available (TABLE 1-2). For instance, the common AF transistor (2SC 374) can be replaced with a universal RCA SK3124A or ECG289A. The AF transistor (2N3904) can be replaced with a universal RCA SK3854 or ECG123AP.

Look up the transistor number within the RCA SK series or with Sylvania's ECG series replacement guide books. Most universal solid-state transistors and IC components can be replaced with RCA, GE, Motorola, NTE, Sylvania, Workman, or Zenith replacements. Simply look up the part number and replace it with an universal replacement. Test the new transistor before installing it.

After obtaining the correct replacements, remove the old transistor with iron and solder-wick from the PC board. Remove the mounting screws on power output transistors, then unsolder the emitter and base terminals. Make sure you have the correct terminals in the right PC board holes. Doublecheck the transistor wire terminals and the bottom base diagram (FIG. 1-17). Do not leave the soldering iron on the transistor terminals too long. Place silicon grease between the transistor, insulator, and heatsink before mounting.

Look in the replacement guide for a universal IC replacement when the original one is not available. Handle it with care; the IC cannot be tested before installation. Leave it in the envelope until it's ready to be installed.

Remove the old one with solder-wick and soldering iron. Start down the outside row of contacts and keep the iron on the mesh at all times. Move the mesh

Table 1-2. Universal Transistor Replacement Chart

Preamp npn	AF amp npn
2SC458-SK3124A-ECG289	2SC372-SK3124A-ECG289
2SC536-SK3245	2SC458-SK3124A-ECG289
2SC693-SK3124A-ECG289	2SC536-SK3245-ECG199
2SC732-SK3245-ECG199	2SC644-SK3245-ECG199
2SC900-SK3899	2SC693-SK3124A-ECG289
2SC1000-SK3245-ECG199	2SC828-SK3931-ECG90
2SC1312-SK3899	2SC945-SK3124A-ECG289
2SC1740-SK3122	2SC1571-SK3245-ECG199
2SC1815-SK3124A-ECG289	2SC1740-ECG3122
2SC2320-SK3122	2SC2240-SK3122

Power output npn	Bias oscillator npn
2SA537-SK3122	2SC537-SK3122
2SA634-SK3250-ECG315	2SC711-SK3899
2SC1030-SK3619	2SC1214-SK3124A-ECG289
2SC1096-SK3248	2SC1317-SK3124A-ECG289
2SC1383-SK3849	2SC1627-SK3449
2SC1568-SK9041	

Preamp pnp	AF amp pnp
2SB173-SK3004-ECG102A	2SB175-SK3004-ECG102A
2SB175-SK3004-ECG102A	2SB186-SK3003-ECG102A
2SB348-SK3004-ECG102A	2SB346-SK3004-ECG102A
2SC732-SK3245-ECG102A	2SC348-SK3004-ECG102A

Power amp pnp	Bias osc. pnp
2SB156-SK3007A-ECG102A	2SB75-SK3004-ECG102A
2SB178-SK3004	2SB172-SK3007A
2SB324-SK3007A	2SB186-SK3004-ECG102A
2SB376-SK3007A	2SB187-SK3004-ECG102A
2SB405-SK3004-ECG102A	2SB365-SK3004-ECG102A
2SB415-SK3004-ECG102A	

RCA SK series   ECG SYLVANIA series

down as it picks up solder. Then, go back and make sure each contact is unsoldered from each pin to PC board wiring. Flick the pin with a pocket knife blade or with a small screwdriver. Mark pin 1 on the PC board with felt pen. Lift the defective IC out by prying underneath the component. Be careful not to damage other components nearby or to break the PC wiring.

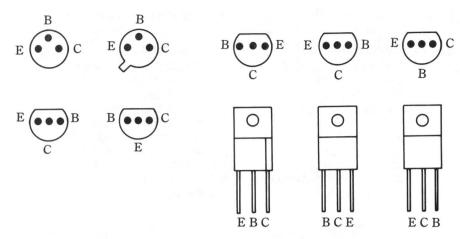

1-17 The many different transistor terminal connections with the AF, driver, and output transistors in the cassette player.

Check for the terminal 1 dot, while the line or indexes determine how to correctly mount the audio IC (FIG. 1-18). Sometimes, only an index indentation is found on the IC. When looking down on top of the IC, pin 1 is to the left of the index. Pin 1 is indicated with a white dot; on other PC boards, the number one and last terminal numbers are marked (FIG. 1-19). Correct mounting of IC is found in the solid-state replacement guide. Place silicon grease and metal heatsink behind large ICs.

1-18 Check for number 1 dot, index, or a white line on top of the transistor to mark and correctly insert the new replacement IC.

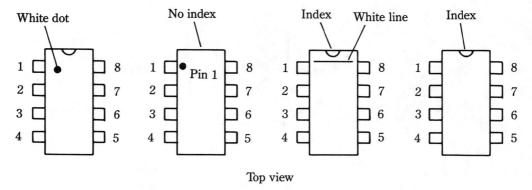

Top view

1-19 Locating pin 1 by looking down on top of the IC.

Silicon and germanium diodes are found in many cassettes' circuits. These diodes can be checked with a VOM, DMM, or diode tester. To get a normal diode check, place the positive lead of DMM to the negative (anode) terminal of the diode and the negative probe to the collector terminal. When checking with the VOM for normal reading, place positive (red) probe to collector and the negative (black) lead to anode terminal of diode (FIG. 1-20).

1. Set the function/range switch to the *K1* diode position on B&K 388-HD DMM.

2. Connect the red test lead to the V-Q-Hz jack and the black lead to the COM jack.

3. To check the forward voltage (VF), connect the red test lead to the anode and the black test lead to the cathode of the diode.

4. The display indicates the forward voltage if diode is normal. Normal diode voltages are approximately 0.3 V for germanium diodes, 0.7 V for silicon diodes, and 1.6 V for LEDs. An overrange (1) indicates an open diode. A shorted diode reads 0 V.

5. To check reverse voltage, reverse the test-lead connections to the diode. The reading should be the same as with open tests leads (1) overrange. Lower readings indicate a leaky diode.

The normal diode will show a reading in one direction and a shorted or leaky diode will read with reverse test leads in both directions. The open diode will not read in any direction. Most defective diodes are leaky or shorted.

Bridge rectifiers can be checked in the very same manner, connect the positive (red) lead to an ac terminal and negative (black) lead to a positive terminal of the bridge rectifier for a normal measurement (FIG. 1-21). The shorted or leaky bridge rectifier can show a low reading in both directions of two or more terminals. Replace the bridge rectifier when only one diode is shorted. The defective bridge diode can be replaced with four separate diodes if one is not available (FIG. 1-22).

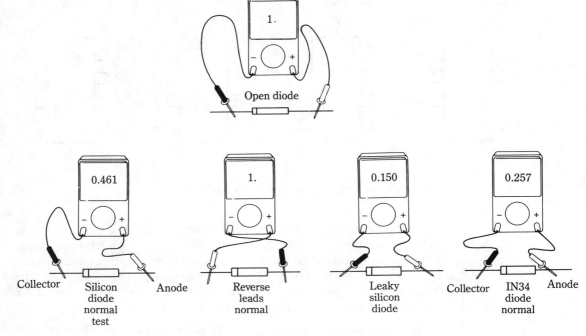

I-20 Checking diodes with the DMM.

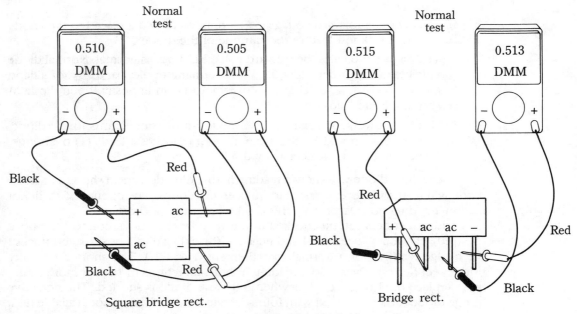

I-2I Checking the diodes in the bridge rectifier.

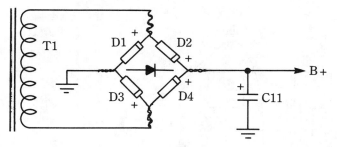

I-22 Replace the bridge rectifier with four I- or 2.5-A diodes when the bridge rectifier is not handy. You might find one arrow on the bridge rectifier in some Japanese circuits.

TROUBLESHOOTING WITHOUT THE EXACT SCHEMATIC

When the exact service manual or schematic is not available, use another manufacturer's schematic diagram. Although the substitute is not exactly the same, it will give you different test points to troubleshoot the circuit. Signal tracing audio circuits can be checked by starting at the tape head winding and going from base to base of each transistor. The output stages can be checked by starting at the volume control.

Usually, cassette audio stereo circuits are laid out on the PC board with the left channel on the left and the right channel on the right. You can start at the speaker and trace the circuit back to locate the output transistors or ICs. Some of these power output transistors and ICs are located on heatsinks. The supply lead from the IC or transistor can be traced back to the power supply. The power supply can be identified with large filter capacitors, diodes, and bridge rectifiers. The suspected part can be checked against the identical one found in the good stereo channel.

Transistor terminals can be identified by letters on the PC board, transistor markings, or with a transistor tester. You can locate the transistor leads with the *diode junction test* of the DMM (FIG. 1-23). Remember, the positive lead of DMM will have a common measurement if placed on the base terminal of an npn transistor. The pnp transistor must have the black lead as the common base terminal. Locate the base terminal by getting two separate tests between the base and collector, and the base and emitter. The npn collector terminal goes to a higher positive voltage, the emitter voltage is very low, and the emitter resistor goes to ground. The pnp collector lead is negative and the emitter has a higher positive voltage. Remember, a leaky transistor has a very low reading with reverse test leads. Also, check the numbers listed on transistor and compare them with the solid-state universal replacement guide.

SIGNAL TRACING WITH A CASSETTE

A weak, dead, or lost signal within the cassette player can be signal traced with a test cassette or one with music and the audio signal tracer or external amplifier. When the signal stops, the defective stage is nearby. Then, take critical voltage and resistance measurements.

npn

npn   npn

EBC

E B C

Marked
on
transistor
body

Marked
on
PC board

npn

E

C

0.1 kΩ

DMM

−   +

Emitter
resistance to
ground

Gnd

6.7 V

DMM

−   +

Collector
voltage
check

1-23 Identify the correct transistor terminals with the DMM when a schematic is not readily available.

Insert a test tape and check the audio signal at the volume control. If the signal is normal here, go stage by stage from the volume control to the speaker. When the signal is weak or when no signal is at the volume control, start at the tape head and check from base to base of each transistor (FIG. 1-24). Check the signal in and out of IC terminals with amplifier or signal tracer.

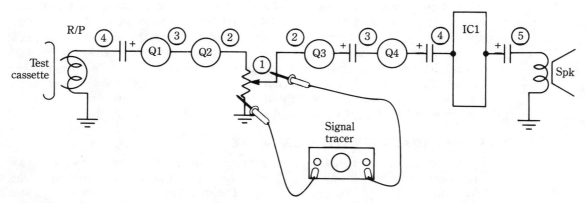

1-24 Start at the volume control with the signal tracer and proceed either way to check for signal loss.

BUILD YOUR OWN TEST EQUIPMENT

Besides the DMM or VOM, several homemade test instruments can speed up the cassette player repairs. The sine-/square-wave generator, 1-kHz audio oscillator, audio signal tracer, and speaker load and noise generator are small test instruments that you can build (FIG. 1-25). The noise generator and sine-/square-wave generator can locate a dead, weak, or distorted audio stage. Use the signal tracer with the homemade 1-kHz test cassette to signal trace the audio circuits. Connect a speaker lead to the speaker connections while working on the audio circuits. Most parts can be picked up at Radio Shack, unless noted after each component.

1-25 Build one or two of these home-constructed testers to help locate defective components within the cassette player.

Sine-/square-wave generator

The sine-/square-wave generator has a frequency range of 20 Hz to 20 kHz and is built around an IC8038 function IC. The power supply is ac operated with a step-down power transformer and an IC regulator. R1 controls the frequency. Build the generator on a perf board and enclose it in a 6-×-6½-×-2¼-inch plastic or metal cabinet (FIG. 1-26). This generator can make many tests within the audio and speed circuits.

Parts list

IC1	7815 15-V regulator.
IC2	8038 function generator, D.C. Electronics, P.O. Box 3203, Scottsdale, AZ 85271-3203.
C1, C3, C4, C7, C9	0.1-μF 50-V ceramic capacitor.
C2	0.0047-μF 50-V monolithic or high-Q ceramic disk capacitor.
C5	0.01-μF 500-V ceramic capacitor.
C6	2200-μF 35-V electrolytic capacitor.

C8	1-μF 35-V electrolytic capacitor.
R1, R10	10-kΩ linear control.
R2	20-kΩ 0.5-W resistor.
R3	8.2-MΩ 0.5-W resistor.
R4, R6	4.7-kΩ 0.5-W resistor.
R5	1-kΩ trimmer or thumb-variable resistor.
R7	3.3-kΩ 0.5-W resistor.
R8, R9	100-kΩ trimmer screwdriver or thumb-variable resistor.
N1	120-Vac neon indicator, 272-704 or equiv.
T1	12.6-Vac 450-mA stepdown transformer, 273-1365 or equiv.
D1	1-A bridge rectifier.
D2	1N914 switching diode.
J1, J2	Banana jacks.
Cabinet	MB-1C beige instrument enclosures, All Electronics, Box 567, Van Nuys, CA 91408.
Perf board	3×4.5.
SW1	On back of R1.
Misc.	ac cord, grommet, hookup wire, bolts, nuts, etc.

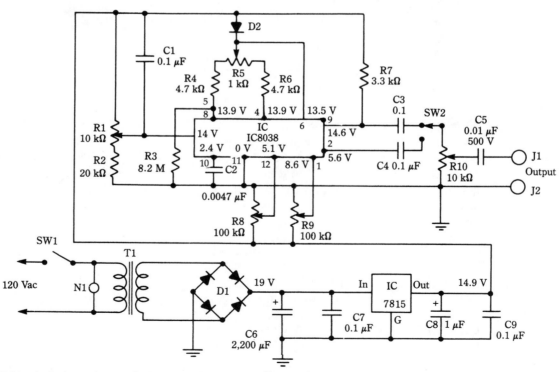

I-26 A diagram of a sine-/square-wave generator operated from the ac power line. This tester can be used for signal tracing, locating distorted stages, head aligning, and correcting speed problems.

IC audio signal tracer

The audio signal tracer can pick up the audio signal after the first AF or predriver stage and trace the signal up to the speaker. This audio tracer consists of only 1 IC and 4-inch 8-Ω speaker. Make sure that C1 has a working voltage of 100 V to prevent damage to the signal tracer IC. The signal tracer is operated from a 9-V battery. Place the components in a large enough cabinet to take a 4-inch PM speaker (FIG. 1-27). Use the audio signal tracer as you would signal trace with a cassette.

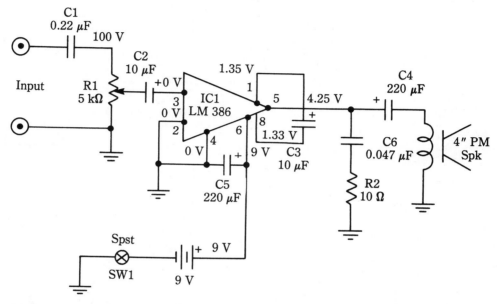

1-27 A simple audio signal tracer used in conjunction with a cassette to locate the defective stage.

Parts list

C1	0.22-μF 100-V ceramic capacitor.
C2, C3	10-μF 35-V electrolytic capacitor.
C4, C5	220-μF 35-V electrolytic capacitor.
C6	0.047-μF 50-V ceramic capacitor.
R1	5-kΩ volume control and switch.
R2	10-Ω 0.5-W resistor.
IC1	LM 386 audio amp, 276-1731 or equiv.
Spk	4-inch 8-Ω pm speaker.
Cabinet	Large enough for 4-inch speakers.
Perf board	Multipurpose board, 276-150 or equiv.
Batt.	9-V battery.
Misc.	Battery cable, alligator clips, hook-up wire, 4-pin IC socket, nuts, and bolts.

White noise generator

The noise generator can be used in signal tracing audio, RF, and IF circuits. Inject the output noise generator to the various audio stages with the ground terminal to common ground of the amplifier. This white noise generator uses a low-priced transistor and op amp IC with a regulated power supply (FIG. 1-28). Start at the volume control or tape head and use the audio amp speaker as an indicator.

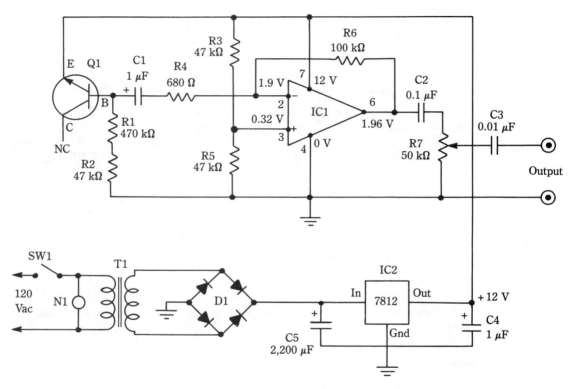

1-28   The white-noise generator can be used to signal trace the AF, RF, and IF circuits of the cassette player.

Parts list

Q1	MPS 2222A npn or ECG123.
IC1	741 op amp.
IC2	7812 12-V regulator.
C1, C4	1-μF 50-V electrolytic capacitor.
C2	0.1-μF 50-V ceramic capacitor.
C3	0.01-μF 450-V ceramic capacitor.
C5	2200-μF 35-V electrolytic capacitor.
R1	470-kΩ 0.5-W resistor.

R2, R3, R5	47-kΩ 0.5-W resistor.
R4	680-Ω 0.5-W resistor.
R6	100-kΩ 0.5-W resistor.
R7	50-kΩ linear control with SPST switch.
D1	1-A bridge rectifier.
T1	300-mA 12-V secondary transformer, 273-1358A or equiv.
Perf board	2.83×1.85, 276-149 or equiv.
SW1	SPST switch on rear of R7.
Case	Plastic box, 3×6×2.
Misc.	ac cord, hookwire, grommet, 8 pm IC socket, etc.

1-kHz audio generator

This little 1-kHz audio generator can be used to inject signal into the various audio stages with the speaker amp as the indicator. Also, the generator can be clipped to the tape head connections to record a 1-kHz test cassette. The signal generator has only a few components which are battery operated and built around the low-priced LM3909 IC (FIG. 1-29). Start at the tape head and inject the 1-kHz audio signal and go from base to base of each audio transistor or to the input terminal of the preamp and power-amp IC.

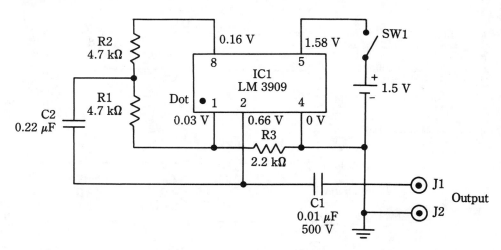

1-29 The 1-kHz audio oscillator can be injected for signal tracing or making that 1-kHz test tape.

Parts list

IC1	LM3909 LED flasher-oscillator IC.
C1	0.01-μF 500-V ceramic capacitor.
C2	0.22-μF 50-V capacitor.
R1, R2	4.7-kΩ 0.5-W resistor.
R3	2.2-kΩ 0.5-W resistor.

Batt.	1.5-V alkaline AA battery.
SW1	Sub-mini slide switch, 275-409 or equiv.
Case	Plastic box, $4 \times 2^{1}/8 \times 1^{5}/8$, 270-231 or equiv.
Perf board	1×2 (cut from larger piece).
J1, J2	Banana jacks.
Misc.	8-pin IC socket, 4/40 bolts and nuts, battery holder, hookwire, etc.

Speaker load

The audio output stage must be loaded down at all times to prevent damage to the output transistors or IC components. In directly coupled output circuits, a defective transistor can damage the PM speakers. The speaker load should be large enough to withstand the wattage of each stereo channel. Remove the speakers and connect the speaker loads to each channel while repairing the amplifier (FIG. 1-30). For low-wattage amplifiers under 20 W, use the 8-Ω load. For higher wattage amplifiers, switch in another 8-Ω 20-W resistor. Place a couple of extra jacks in the circuit for amplifier voltage adjustment.

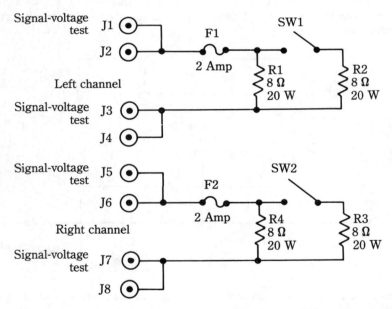

1-30 The dual-speaker load prevents speaker and amplifier damage while you service the audio stages of the cassette player.

Parts list

F1, F2	2-A fuses.
R1, R2, R3, R4	8-Ω 20-W resistor.
J1 through J8	Banana jacks.

SW1, SW2	Toggle switches.
Cabinet	Plastic economy box.
Misc.	Terminal strips, hookup wire, nuts, bolts, etc.

HEAD AZIMUTH AND CURRENT TESTS

The tape head must be properly lined horizontally for optimum sound reproduction. Usually, one side of the tape head is fastened with a small screw and the other side with adjustable spring (FIG. 1-31). Improper head azimuth adjustment can cause distortion and loss of high frequencies. You can adjust the azimuth screw by playing recorded cassette of violins or high-pitched music to maximum into speakers.

Head
azimuth
screw

1-31 The head azimuth adjusts the tape head horizontally with the tape, The azimuth screw is located alongside of the tape head, near the tension spring.

Accurate azimuth adjustment can be made with a 3-, 6.3-, or 10-kHz test cassette with 8-Ω load instead of speakers. Connect the low ac range of DMM across the 8-Ω resistor or use the dummy speaker load project (FIG. 1-32). Play the recorded cassette and adjust the azimuth screw for maximum on the DMM.

Current adjustment

To make sure that bias voltage and the bias oscillator is operating, take a current test at the R/P tape head. Bias adjustment can be made at the same time. Insert a

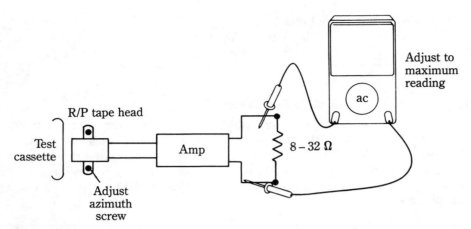

1-32 Connect the frequency counter of the DMM across the 8- to 32-Ω resistor at the speaker or headphone output jack.

100-Ω resistor between the ground (shielded) lead of the tape head and shield (FIG. 1-33). Place cassette player in the record mode for this adjustment. Measure the voltage across the resistor with a VTVM or DMM. Most tape head current runs between 20 and 65 mV. Adjust the variable bias resistors at the tape head for correct voltages at each channel.

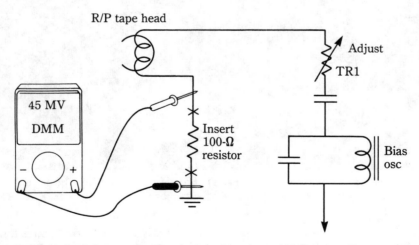

1-33 Adjust and check for tape-head current by inserting a 100-Ω resistor between the shield and ground of the tape head (20 to 65 mV).

TAPE HEAD CLEANING

The tape head should be cleaned at least six times per year if the cassette player is in constant usage. Some manufacturers recommend cleaning after 40 hours of operation. Iron oxide particles from the magnetic tape will build up on all components that come in direct contact with the tape. This excessive oxide can produce garbled or muffled sound during playback. One channel gap can be closed with oxide, which results in no sound or recording from that channel. Oxide deposits can cause an improper erase function and prevent automatic stop operations. Keep magnetic metal away from the tape head at all times.

Clean the tape and erase heads with cleaning stick and alcohol. This action can be done with the cassette door open or off (FIG. 1-34). Make sure that all packed oxide is removed from the tape head area. Wipe off the capstan drive shaft and pressure roller. Apply heavy pressure to clean the rubber pressure roller. Rotate the roller as it is cleaned and leave a black rubber surface.

1-34 The erase head is mounted before R/P tape head to erase previous recording. Clean off all oxide with a cleaning stick and alcohol.

Another convenient method is to use a cassette head cleaning tape. Some are dry while others apply wet cleaning solution. Do not play the cleaning cassette too long. Follow the manufacturer's directions. Of course, a good clean up with alcohol and cleaning stick is best.

Always clean around the tape area when the cassette player has been repaired. Now is the time to get down inside the cassette rotation area for good clean up. Besides cleaning tape heads and the pinch roller, touch up the capstan, flywheel, idlers, and turntables. Keeping oxide dust from building on tape path components can prevent future cassette player repairs.

TAPE AND ERASE HEAD PROBLEMS

Besides collecting oxide dust, the R/P tape head can become open, intermittent, or cause distortion in the speakers. The worn tape head can cause a loss of high frequencies. A magnetized tape head can cause extra noise in recording. An open head winding will produce a dead channel. Poorly soldered connections or broken internal connections can cause intermittent music. Check the open or high-resistance head winding with the ohmmeter (TABLE 1-3). Inspect the cable wires and resolder for intermittent conditions. Make sure that one screw is not loose, which would let the head swing out of line.

Table 1-3. Typical Tape Head Resistance

Model	Tape head resistance
GE-3-54808KA	225 Ω Actual measurements
Panasonic RQ-L315	315 Ω Actual measurements
Sony M440V	348 Ω Actual measurements
Sony TCS-430	512 Ω Actual measurements

Typical R/P tape head resistance 200 to 830 Ω
Typical erase head resistance 200 to 1000 Ω

Erase head

The erase head is mounted ahead of the P/R head so that it will erase any previous recording. Usually, the erase head has only two leads and is excited by a dc voltage or by the bias oscillator circuit. Suspect the erase head when garbled or two different recordings are heard in the speakers. Check for open erase head with the low ohmmeter range (FIG. 1-35). Make sure that the tape is pressed against erase head when operating. The erase head resistance can vary from 200 to 1000 Ω.

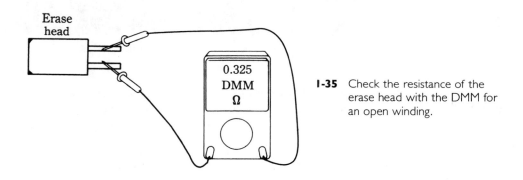

1-35 Check the resistance of the erase head with the DMM for an open winding.

DEMAGNETIZE TAPE HEADS

The magnetized tape head can produce background noise within the cassette recording. Hissing noise in playback can result from a magnetized head. Keep tools that are magnetized away from the tape heads. Magnetized screwdrivers should not be placed near the cassette player. Do not set the cassette player near or on top of large speaker columns.

The tape head should be demagnetized at least twice per year. Insert an ordinary cassette demagnetizer, which looks like and loads like any cassette. The demagnetizer with probe can get down inside the cassette loading area to demagnetize the tape heads. Do not shut the demagnetizer off when it is next to the tape head. Pull the unit out, then shut off the demagnetizer. Head demagnetizer and head cleaning kits can be found at most electronics stores.

SPEED ADJUSTMENTS

The speed of the cassette player can be checked with a test tape and a frequency counter (FIG. 1-36). Insert a 1- or a 3-kHz test cassette and place it in the play mode. Connect a 10-Ω resistor at the earphone jack. Measure the frequency at the DMM frequency meter. If the reading is at 1 kHz, the speed is correct. A higher reading indicates faster speed and a lower measurement indicates a slower speed. Don't worry if reading is around 1 kHz. Some larger cassette players have regulated speed adjustments or you can find a speed adjustment in the end bell of the dc motor. Slow speeds can be caused by dirty, oily, or loose belts. Oil deposits on the capstan/flywheel can produce slow speeds.

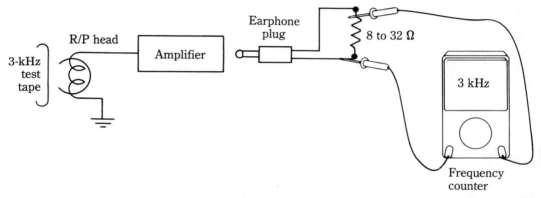

1-36 Check the speed of the cassette player with a 1- or 3-kHz test cassette and frequency counter of the DMM.

Yes, with a few hand tools, a DMM or a VOM, and one or two home-constructed test instruments, you can repair that broken cassette player. Remember, the sine-/square-wave project can be used for signal injection, distortion location, head azimuth adjustment, and speed tests. Just pick out the chapter that relates to the cassette player you are servicing and get started. Of course, if you read each chapter, many repairs are related to each cassette player.

Chapter **2**

Repairing tag-along personal cassette players

*L*ow-priced tag-along cassette players are used during physical activities or just for listening to music. Some of these units are only players and have no recording features, although some do record and play (FIG. 2-1). You can find some with only AM radio or AM/FM radio combined with cassette player. The stereo circuits are the same as those found in chapter 3. These mini-cassette players or recorders fit into the shirt or coat pocket, snap on a belt, or have their own carrying strap.

The cassette player might operate only from earphones with no enclosed speaker (FIG. 2-2). Some units come with self-enclosed earphones while others have to be supplied. These earphone sets can be picked up anywhere with 8 to 40 Ω impedance. You can choose from those that clamp over the head to ones that are small enough to fit inside the ear (FIG. 2-3).

Both speakers and earphone operation are used in some of the tag-along mini-cassette players. Of course, the speaker is disconnected when earphones are inserted. The manual speaker is usually located at the bottom of the cassette players.

PLAYER ONLY

The lower priced tag-along cassette player can be used for playing only, without any recording features. Here, only one small tape head has direct belt-drive features. Stop, fast forward, and play are the only pushbuttons, which makes this unit easier to operate (FIG. 2-4).

Only a motor belt, motor, and capstan/flywheel move. The tape motor rotates in only one direction and drives the flywheel in the same direction. The motor belt can be a square- or flat-type drive belt (FIG. 2-5). Very few speed problems are found with this type of drive system. Simply clean the motor pulley, drive belt, and flywheel for slow speed problems.

33

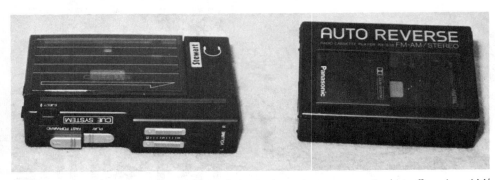

2-1   The personal cassette player might only have play, auto reverse, auto shut off, and an AM/FM radio in the same unit.

2-2   Some cassette players have no speakers, only earphone jacks.

2-3 Many low-priced and expensive earphones are available. Some clamp over the ears and others can be placed inside the ear.

2-4 Only one R/P tape head and take-up operating reel are in the play-only cassette model.

2-5   The tape plays in only one direction with a flat- or square-motor belt in play-only cassette models.

CASSETTE TAPES

The mini or tag-along cassette player/recorder uses normal tape cassettes. Use C-90 and not C-120 tapes in these machines. The tape is easily broken or stretched; if not used with extreme care, it can get tangled with the capstan or pressure roller. Cassette tapes, both recorded and unrecorded, should not be stored in locations with high temperatures, high humidity, or direct sunlight. Never place a recorded cassette near a magnetic source, such as a magnet, large stereo speakers, or a TV set.

Do not use a cracked plastic cassette in the tape player. A defective cassette can run slow, have sound dropouts, and crimped tape. Try another cassette when the player begins to drag or slow down and compare them. The defective cassette might have poor recording qualities with worn conditions and poor high-frequency response. Some cassettes can quickly spill out of the cassette and wrap around the pinch roller and capstan. Do not forget to check the tape head for magnetization.

VAS, VOX, OR VOR SYSTEMS

In the higher priced mini-cassette recorders, you can find a *VAS (Voice Activated System)* system. When recording using the VAS function, the sound is recorded automatically so that no tape is wasted. With the VAS/pause switch, the tape runs when sound is picked up by the built-in or external microphone. Then, when no sound is picked up, tape stops running automatically (about 3 or 4 seconds later).

The volume/VAS level control should be normally set to the (4 to 7) position. To record loud sound only, rotate the control toward (1 to 3). To record low sounds, rotate toward (8 to 10). No sound is recorded at the 0 position. Remember to check the position of volume/VAS level control when the player will not record or start rotating when in VAS operation (FIG. 2-6). The recording level is automatically adjusted, regardless of the position of the volume/VAS control level control.

2-6 The VAS recording switch and level control in the Panasonic RQ-L315 recorder.

VAS recording with Panasonic's RQ-L315 cassette player has a VAS, off, and a pause switch. When switch is off, the player must be turned on by hitting the play or record switch. The VAS/pause switch can be used to stop the tape movement temporarily during recording or playback. Do not use the VAS/pause switch to stop the tape for a long period of time. Remember, the unit is not turned off when the VAS/pause switch is set to pause or on. Always use the stop button to turn off the unit.

HEAD AND CABINET CLEANING

The sound quality of the cassette player might become weak and distorted with a dirty tape head. Remember, the tape head, capstan, and pressure rollers are in contact with the tape at all times. Make it a habit to clean them after 10 hours of playing (FIG. 2-7).

Clean the dirty tape head with alcohol and a cleaning stick or use dampened cloth with a little alcohol. Manufacturers rarely recommend using cleaning tapes because some contain abrasives and can cause premature head wear. Just clean them with alcohol and a cleaning stick.

Do not clean the plastic cabinet with paint thinner or benzene. Clean the cabinet with a cloth dampened in a mild soap-and-water solution. Avoid excessive moisture. Wipe the cabinet dry with a soft cloth. Try to avoid spray-type cleaners when the cassette mechanism is in the cabinet, because chemicals can discolor plastic body.

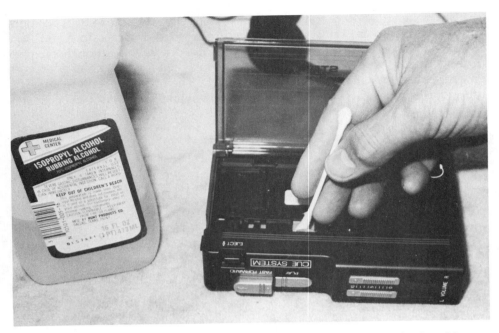

2-7   Clean the tape head with a high grade of isopropyl alcohol and a cleaning stick.

HEAD DEMAGNETIZATION

Use a cassette tape demagnetizer tool. Several different demagnetizer tools can be found at most electronic supply stores. Follow carefully the instructions that are supplied with the device. Do not bring any metal objects or tools that are magnetized near the tape head.

THE CASSETTE MECHANISM

The topside of the cassette mechanism (FIG. 2-8) consists of supply and take up reels, the capstan, R/P head, and erase head. The bottom view of the cassette mechanism can show the small drive motor, drive belt, capstan/flywheel, and various idler wheels. The motor rotates the capstan/flywheel with a rubber flat or square belt (FIG. 2-9). The idler wheels and arm are placed into the system by push-buttons.

When the fast forward button is pressed, the take-up reel is pulled by an idler pulley, engaged against the flywheel. Often, the pinch roller is not engaged or the capstan is on the tape in fast forward mode. When pressed, the play/record button applies voltage to the motor, rotates the belt, and the flywheel/capstan. The capstan with pinch roller pulls the tape from the supply reel and the slack is taken up by the take-up reel. You must push both the play and record buttons to set the unit into the record mode (FIG. 2-10).

2-8 The top view of the cassette mechanism in the Panasonic mini-cassette recorder.

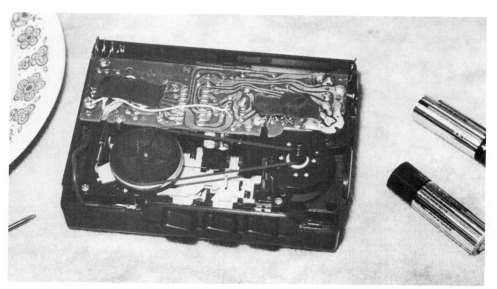

2-9 The motor pulley rotates the drive belt, which turns the capstan/flywheel for tape action.

SLOW SPEED

Slow speed can be caused by a dirty motor pulley or oil spots upon pulley. A binding or dry capstan/flywheel bearing can slow down the cassette player. A worn, stretched, or oily belt can cause slower speeds (FIG. 2-11). Clean the packed oxide

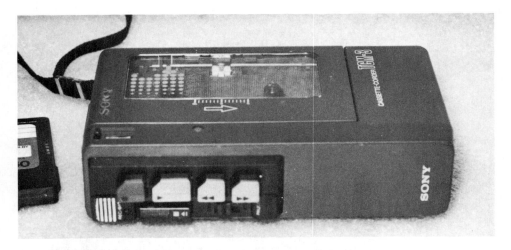

2-10   You must push both play and record buttons to record in most cassette players, including Sony's T6M-3 player.

2-11   Check the belt for loose, dirty, oily, and cracked areas.

tape head and the dry pinch roller. Slower speeds can result if excessive tape is wrapped around the pinch roller and in between the rubber roller and the bearing plate. Suspect a defective motor for erratic or slow speeds after the rotating parts are clean. Do not overlook a defective cassette, try another one.

DISASSEMBLING THE COVERS

Simply removing the bottom screws lets the cover come off. Besides the bottom screws one or two top screws inside the top lid must be removed to drop off the bottom covers in a GE 3-54808KA cassette player. Remove the mechanism with the PC board by removing two white chassis screws (FIG. 2-12). Now, the tape heads, capstans, and the motor belt can be cleaned easier.

2-12 Remove 6 screws to remove bottom cover and PC board in the GE 3-54808KA cassette player.

Notice that within this cassette player only the take-up reel rotates. The supply reel is nothing more than a plastic post. The motor drive belt drives the plastic-metal flywheel. At the top of the flywheel, under the capstan, a pulley drives another small belt that rotates the take-up reel (FIG. 2-13). Here, the fast forward speed is the same as the speed of the play take-up spindle, except that the capstan does not engage the tape. When replacing the PC board and mechanism assembly, make sure that the small wires are in the original position.

NO SOUND/NO TAPE MOVEMENT

Check for dead or weak batteries. Replace weak batteries. If the tape still does not rotate, inspect the leaf switch. Often, low-priced cassette players have a small leaf switch that makes contact when play, record, fast forward, and rewind features are used. Clean the copper spring-type contacts with cleaning fluid and a stick (FIG. 2-14). Sometimes, pulling a piece of thin cardboard (match cover) through closed switched contacts cleans them. Suspect a defective motor or motor circuit when the batteries and the leaf switch are normal.

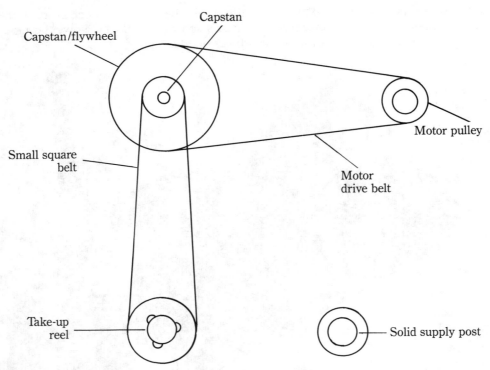

2-13 A motor belt drives the capstan/flywheel from the motor pulley and the take-up reel assembly is belt driven from a small pulley on the hub of the capstan.

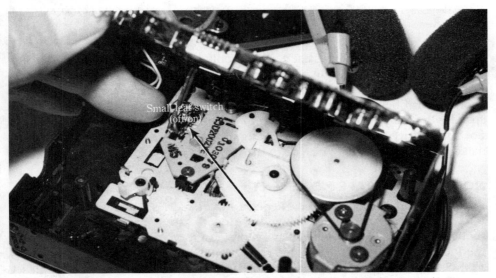

2-14 Small leaf-switch contacts are pressed together in the various modes to provide voltage for the motor and amplifier.

NO FAST FORWARD

In most surface-drive tape mechanisms, the idler wheel is shoved over to rotate the take-up reel. The idler wheel is rotated by friction drive against a wheel that is attached to the capstan/flywheel shaft. If the player operates normally in play and slow in fast forward, suspect slippage on the idler drive area (FIG. 2-15). Clean all drive surfaces. When the fast forward is belt driven, clean the belt and drive pulley. If both play and fast forward are slow, clean the motor belt and flywheel surfaces.

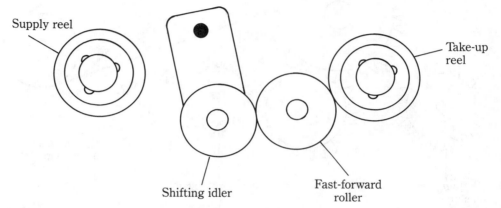

Supply reel

Take-up reel

Shifting idler

Fast-forward roller

2-15   The idler wheel is shifted toward the fast-forward roller. It then drives the take-up reel at a faster speed.

Fast forward and rewind in the Panasonic RQ-L315 recorder are actually driven from small plastic gears. The small white plastic teeth mesh when switched to fast forward. The capstan gear rotates a large idler wheel and drives another shifting idler gear wheel (FIG. 2-16). The idler gear wheel is shifted toward the take-up spindle, which engages two small gear wheels. At the bottom of the take-up reel is a plastic gear wheel that rotates in fast forward and play.

These gear-type assemblies rarely lose speed or slip while rotating. Check for broken gear teeth or jammed gears when fast forward does not rotate. A missing *C* washer can let the small gears fall out of line and disable fast forward and play.

POOR REWIND

Rewind and fast forward run faster than the play/record. In older and lower priced players in rewind mode the shifting idler wheel is shifted when rewind button is pushed against the turntable reel assembly (FIG. 2-17). Check for worn or slick surfaces on idler or turntable drive areas. Clean them with alcohol. Remember, the pinch roller does not rotate in either rewind or fast forward.

With geared drive systems, the idler is shifted against the gear of the supply spindle. Usually, the rewind speed is lower than fast forward (FIG. 2-18). In rewind, capstan gear rotates the large drive gear, which in turn rotates the shifting idler gear, and the idler drives the gear on the bottom of the supply spindle.

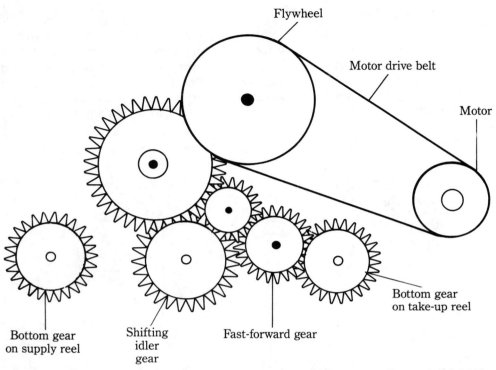

2-16   Small plastic gears are shifted into position to operate functions in the Panasonic RQ-L315 recorder. Here, the idler gear is pressed against two small gears that drive the take-up spindle at a faster fast-forward speed.

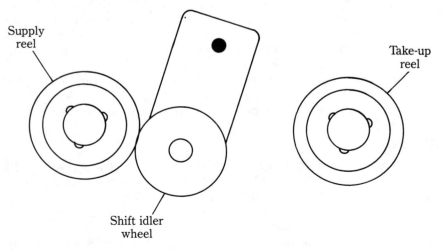

2-17   The idler wheel shifts toward the supply reel in the rewind mode.

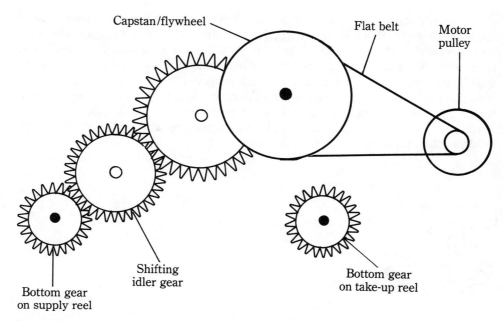

Capstan/flywheel

Flat belt

Motor pulley

Bottom gear on supply reel

Shifting idler gear

Bottom gear on take-up reel

2-18 The shifting idler gear presses against the gear, which is attached to the rewind spindle for rewind operation.

NO AUTOMATIC SHUT OFF

Excessive tension of the tape engages and triggers a small ejection lever that mechanically releases the play/record assembly and shuts off tape rotation. While in the larger units, mechanical and electronic automatic shutoff systems are found. Sometimes the ejection lever is called a detection or contact piece (FIG. 2-19).

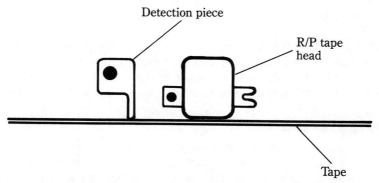

Detection piece

R/P tape head

Tape

2-19 The eject piece or detection piece is mounted close to the tape path to shutdown the player when tape end is reached, in automatic shutdown.

The automatic stop-eject or detection piece has a plastic cover over a metal angle mechanism, that can have adjustment at the end where it triggers the play/ record assembly, the automatic stop. The ejection piece is mounted alongside the tape head. When the end of the tape has been reached, the tape exerts pressure against the ejection piece and mechanically triggers the play/record mechanism.

The end of the eject lever contains a notched metal end that can be widened for understroke and pinched together with a pair of long-nose pliers for over-stroke (FIG. 2-20). The adjustment of closing or opening notched lever should be made so the mechanism acts when the tape ends.

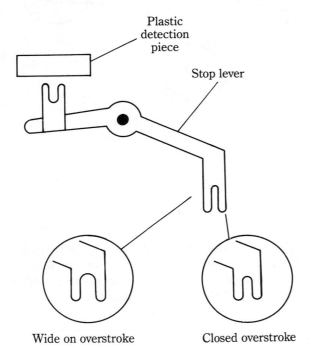

Plastic detection piece

Stop lever

Wide on overstroke   Closed overstroke

2-20 Adjustment of the detection shutdown lever can be spread apart or squeezed together to make tape player shut off at the end of the tape.

Check the adjustment of levers when the tape will not shut off mechanically. Notice if the lever is bent out of line. Does the eject or detection piece ride against the tape at the end? Straighten up the lever or replace it for auto shut-off. Place a drop of oil at the bearing if the ejection piece is binding or difficult to move.

BELT DRIVE SYSTEMS

You can find several belt drive systems within the cassette mechanism. Most have a motor drive belt to the capstan assembly (FIG. 2-21). The drive belt is very small in micro- or mini-cassette players. The motor drive belt in the Panasonic RQL315 model is only two inches long. The belt can be flat or square.

Besides the motor drive belt, another belt runs from the flywheel to the take-up reel. Some mini-cassette players have a fast-forward drive belt. Because these

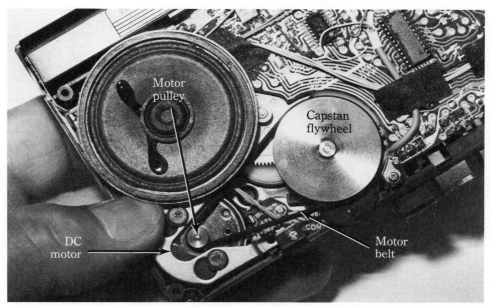

2-21 Only a two-inch, thin square rubber belt drives the Panasonic cassette player.

belts are very small and thin, they have a tendency to stretch and cause slow speeds. Clean each belt with speed problems with alcohol and a cloth. After clean up, if the speed is still abnormal, replace the motor drive belt.

CASSETTE SWITCHES

Many small switches are in the personal cassette recorder. The *sound-level equalizer (SLE)* switch improves recording in locations away from the source—especially in class and conference rooms. The pause and VAS are slide switches. Usually, the radio-tape switch is a slide switch. When these functions do not work or are erratic, spray cleaning fluid down into the switch area. Use it sparingly so that you don't spill fluid over the belts and idler wheels.

The on/off switch that furnishes power to the motor and amp circuits might be a leaf switch and that is pressed on when record, play, rewind, or fast forward are used. The small switch contacts might become dirty, but you can clean them with cleaning fluid and a stick. Suspect a defective or dirty leaf switch when unit appears intermittent. This switch is squeezed shut with a metal lever (FIG. 2-22).

The tape-speed switch, VOR and VOX switches can be slide types with several positions. The record/play switch in the Radio Shack VSC-2001 recorder has five separate contacts when switched in record or playback modes. Besides this switch, the earphone, remote and external power jack provide a short-circuit switch when the male plug is out of the circuit (FIG. 2-23). Most defective switch contacts can be cleaned and lined up. Replace them if they are broken.

2-22   The on/off leaf switch in the Panasonic RQ-L315 mini-cassette player.

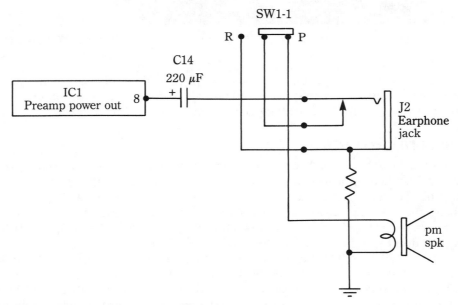

2-23   The earphone jack switches in the speaker terminal when the plug is out of the circuit.

EARLY AUDIO CIRCUITS

In the early cassette solid-state audio circuits, transistors were used throughout. The preamp, AF amplifier, and driver amplifier stages provided good audio to drive the transistorized push-pull output stages (FIG. 2-24). The two transistor outputs were transformer coupled to the PM speaker. A single oscillator transistor furnished bias to the erase and R/P head.

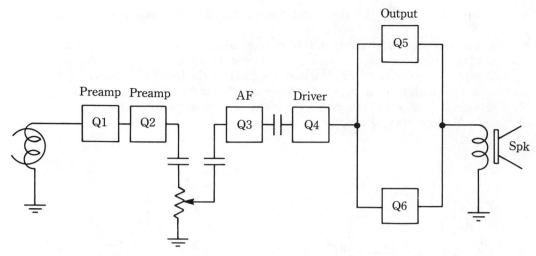

2-24   The block diagram of an early solid-state cassette player with transistors.

The record/playback head is switched into the first audio preamp stage in playback mode. The tape head is capacity coupled (C1) to the base of Q1. The base of Q2 is directly coupled to the collector terminal of Q1. Volume control (VR-1) is tied into the preamp circuit by capacitor C5 (FIG. 2-25).

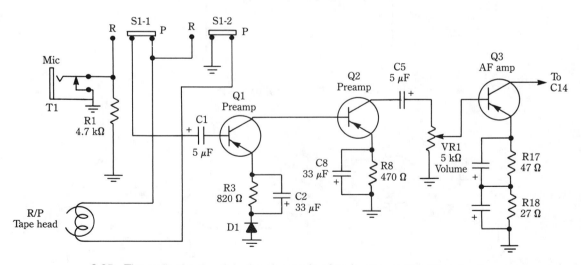

2-25   The audio signal path in the play mode of early preamp transistor circuits.

VR1 controls the tape head and preamp volume that is applied to AF amplifier Q3. The base of Q4 is capacity coupled to the collector of Q3. Both Q5 and Q6 base circuits are transformer coupled to the collector terminal of Q4. The collector terminals of the push-pull output transistors are transformer coupled to the speaker and the earphone jack (FIG. 2-26). Notice that all of the transistors are early pnp types.

Next came transistor preamps and IC power output circuits. The record/playback head is capacity coupled to the transistor base (Q1) and to IC1 through a volume control. The amount of VR1 volume is applied to the input terminals of IC1. C14 capacity couples the audio signal to the speaker and to the earphone in the playback mode (FIG. 2-27). The audio circuit voltage is supplied from four 1.5-V batteries and a power cord adaptor.

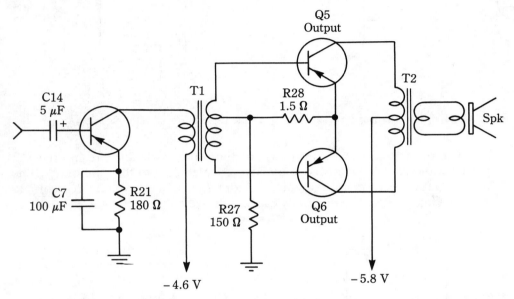

2-26   Driver and push-pull audio-output transistors in a pnp output circuit.

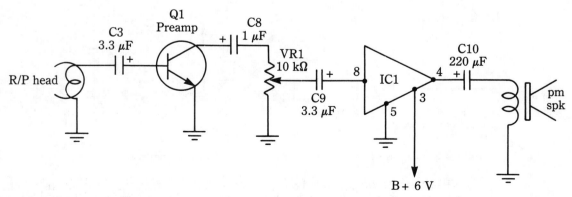

2-27   The simple preamp transistor and typical AF power-output IC audio circuit.

Today, most small audio circuits are found in one or two IC components (FIG. 2-28). The preamp audio stages can have a separate IC from the power output IC. The large IC found in the small cassette player contains all of the audio stages.

The microphone input can be directly connected to the input circuits through C105. The elecret condenser mic that is used in many of today's cassette players has a fixed supply voltage. Often, the voltage is supplied from an RC-filtered network.

2-28 Only one audio IC is used in today's small cassette player.

In the playback mode, the R/P tape head is switched, directly into the circuit with S101, to pin 24 of IC1. The speaker is coupled through R113 to pin 8 of IC1. The earphone audio is taken from pin 10 and R114 (FIG. 2-29). The speaker is disconnected from the circuit when the earphone plug is inserted.

SURFACE-MOUNTED COMPONENTS

Surface-mounted components are found in the higher priced personal cassette recorders. These surface-mounted parts are tiny, except for ICs and solid-state processors. They should not be touched, unless you have the patience and special equipment to remove and install components. The transistors and ICs must be replaced with original components. Some of the small resistors can be replaced with 1/8-scale parts. Surface-mounted components are located on the printed board side and regular capacitors and parts are mounted on the other side of the board (FIG. 2-30).

DEAD PLAYBACK

The tape is rotating with no sound from the speaker. Insert the earphone to see if the speaker is defective. If it is still silent, clip a small PM speaker with an electrolytic coupling capacitor and ground lead to the output circuit (FIG. 2-31). If it

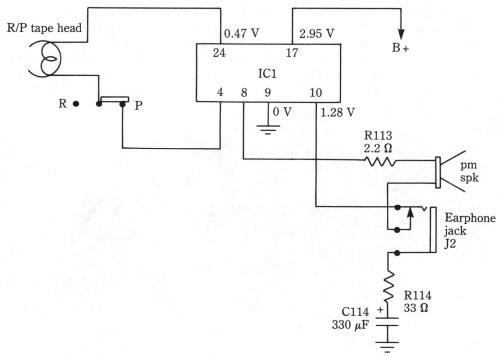

2-29 The signal path from the playback head input and output of the IC to the speaker circuits.

2-30 Surface-mounted components are used on the PC board in the Panasonic cassette recorder.

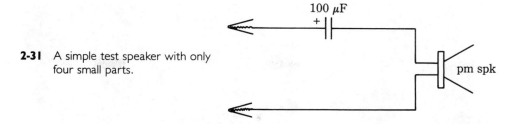

2-31 A simple test speaker with only four small parts.

sounds, check for poor switching contacts in the earphone circuits. Check for broken speaker leads. Sometimes these small single-wire speaker leads get wedged between the PC board and parts, which breaks the leads.

If the audio output stages are normal, insert a recorded cassette with music and signal trace the audio with an outside amp or audio tester. Trace the audio from the playback head to the input terminal of the IC or base of the first preamp transistor. The signal is quite weak at this point. Signal trace the audio from the base to the collector of each transistor stage.

In IC preamps, check the audio at the input and the output terminals of the IC. You can break the audio circuits in half by checking the audio at the volume control. If you find adequate volume at the volume control, the output IC circuits are defective. When you find weak sound or no sound at the volume control, the defective stage is in the input circuits. Do not forget to check the supply voltage of each IC. No voltage or low voltage can indicate a defective power supply or a leaky IC.

DISTORTED SOUND

Muffled or distorted audio is often found in defective audio output stages. Leaky transistors and IC components can cause distorted audio. First, check the input audio that goes into the output stages for distorted sound. If the input is free of distortion, check the audio output stages and the speaker.

Substitute another PM speaker for the enclosed one. Clip the PM speaker, preferably a larger speaker, across the speaker terminals (FIG. 2-32). Disconnect one lead to the built-in speaker. Replace small speaker if the audio is distorted.

Check the supply voltage that is applied to the output IC component. Measure the voltage at all IC terminals and see if any are different from those that are listed on the schematic. Replace the defective IC or transistor if the input sound is normal, all voltages are fairly good, and yet the speaker audio is distorted. Do not overlook the leaky or open electrolytic speaker capacitor for distorted, weak, or intermittent sound (FIG. 2-33).

REMOVING THE SMALL CASSETTE LID OR COVER

On personal cassette players, remove the lid by prying the small plastic hubs out of their sockets. These plastic lids have a tendency to snap off. The lid is quite easy to remove, but it is difficult to replace. After the cassette player is serviced, leave the main mechanism out of the unit until you replace the top lid. Sometimes

2-32 Just clip a small PM speaker across the suspected speaker with one lead removed.

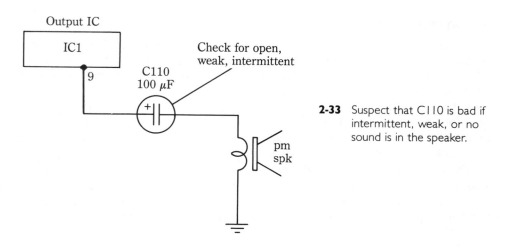

2-33 Suspect that C110 is bad if intermittent, weak, or no sound is in the speaker.

the small spring lock on one end has to be loosened. Loosen these screws, but do not remove them (FIG. 2-34).

Now, place the plastic cover into top case, with leaf spring upward. Close the unit together and snap the other plastic end into the hub area. A miniature screwdriver can help to pry upon plastic end piece. Make sure the top cover will open,

2-34 Replace the plastic cover or lid by snapping the small ends into the hub and spring area.

stay apart, and snap closed after being replaced. Then, tighten the small screws on metal mounting springs.

Replace the mechanism after the top cover is replaced. Be careful not to lose the miniature screws. Place them in a saucer or cap lid. You cannot buy these small screws over the counter. Of course, you can take some from a discarded or defective cassette player.

REPAIRING EARPHONES

Besides cleaning the tape head and replacing small batteries, earphones cause the most problems. The cord breaks at the earphone or where it goes into the male plug and the male plug makes a poor connection at the earphone jack. Any earphones can be used.

The manual phones have only two connections at the male plug while stereo earphones have three wire connections. Also, the stereo earphone jack has a common ground and two ungrounded wires. Use the ohmmeter to make tests at the male plug.

Clip the meter test leads to the common ground (long area at the back) and the outside tip of the pair of stereo earphones (FIG. 2-35). Then, clip the DMM test probe to the other male connection. The resistance should be from 7 to 50 Ω, depending on the headphone impedance. The higher the impedance, the greater the resistance. No measurement indicates an open cord wire or phone winding.

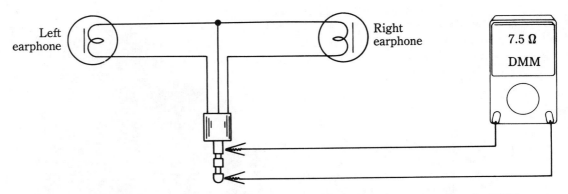

2-35   Check the stereo headphones for continuity with the low ohmmeter range of a DMM.

Flex the phone cord while it is clipped to the meter. When a poor connection is found, the meter hand will rapidly change. Clean the male plug for erratic or intermittent reception. Move the cord near the plug to determine if the breakage occurs at the male plug.

Remove the rubber sponge cover and remove the speaker element (FIG. 2-36). Inspect the cord for breakage. Often, the small flexible wire is pulled out of the earphone soldered connection. Resolder the poor connection for erratic audio. If the wire is broken, tin the flexible wires with solder and solder paste. Resolder the broken wire to the earphone.

2-36   Some cassette players have no speakers, only earphone jacks.

Check the earphone for open internal wiring or winding at the earphone connections. Measure one of the earphone's total resistance. Of course, this resistance is slightly lower than the impedance of the earphones. For instance, the 8-Ω impedance might have a resistance of 7 to 7.5 Ω.

If the cord is broken at the male plug or if the plug is damaged, replace it. Most electronic stores have standard male plug replacements. When soldering male plug connections, keep the wire bare ends short and make a clean soldered connection. Do not let the bare ends or wires touch the outside metal ground cover. Tin the back $1/8$ inch of the cord and solder to the small plug terminals. Place cellophane tape between the connection and the ground lug to keep it from shorting.

Chapter **3**

Servicing personal and portable stereo cassette players

The personal cassette player might contain two separate stereo channels, an AM/FM stereo radio receiver, a cassette recorder (FIG. 3-1), monitor speakers, auto-stop, 4-function cassette transport, auto reverse, Dolby sound, dynamic bass sound, extra bass sound, or dynamic super loudness control.

The personal cassette player can be purchased from $19.95 up to $269.95. The super models contain AM/FM synthesized tuners, stereo recording, Dolby, up to 14 preset stations, mega bass, and quick rechargeable batteries. Most of these personal cassette players and recorders operate from two AA batteries.

STEREO FEATURES

The stereo cassette player might have one or two ICs as preamps and output amplifiers in the audio system. In smaller players, one large IC can contain all of the audio circuits. The volume control can be located between preamp and output IC chips.

Often, the input stereo circuits consist of two different head and external microphone jacks. When the external mic is in the circuit, the internal microphone is switched from the circuit. Both the microphone and R/P tape heads are switched in the right and left input circuits (FIG. 3-2). The left and right R/P tape heads and the erase head are excited from a transistorized oscillator circuit.

3-1   The inside view of the Panasonic RQ-L315 personal cassette recorder with some surface-mounted components.

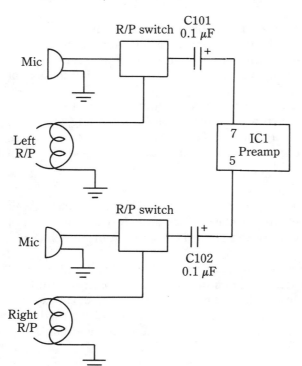

3-2   The stereo audio circuits have identical left and right channels with switching in the R/P tape head and microphone-input circuits.

In play mode, the tape head signal is switched to both left and right channels and amplified by the preamp stages. A dual volume control sets the amount of volume that is applied to the stereo IC power amps. The output audio signal is switched to the left and right speakers (FIG. 3-3).

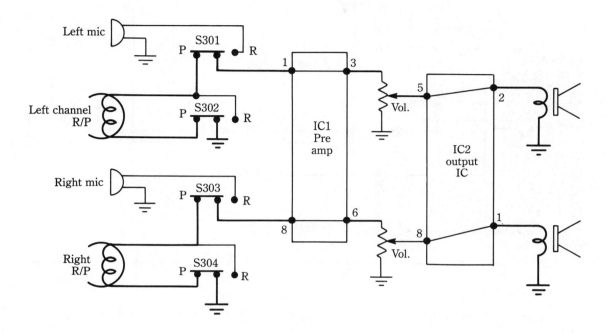

Heavy dark lines
signal path

3-3 The dark lines show the audio signal from the tape heads to the speaker with IC audio components.

When recording, the left and right microphones are switched into the circuit with S300 and applied to the preamp circuits. The audio signal is amplified and passed on to the audio output or to a separate IC record amp and switched back to the R/P tape heads. The audio signal picked up by the microphone is recorded on the tape by passing through the preamp and the audio amp stages (FIG. 3-4).

Also when recording, the erase head is excited by the bias oscillator. At the same time, the bias is applied to each stereo recording tape head for optimum recording. Voltage is only applied to the bias oscillator when recording. The bias oscillator circuit is dead in play mode (FIG. 3-5).

ERRATIC PLAY

Intermittent or erratic play can be caused by a dirty tape head, dirty R/P switch contacts, an intermittent preamp, an intermittent audio output amp, or a bad speaker. Usually, the intermittent or erratic sound will only occur in one channel.

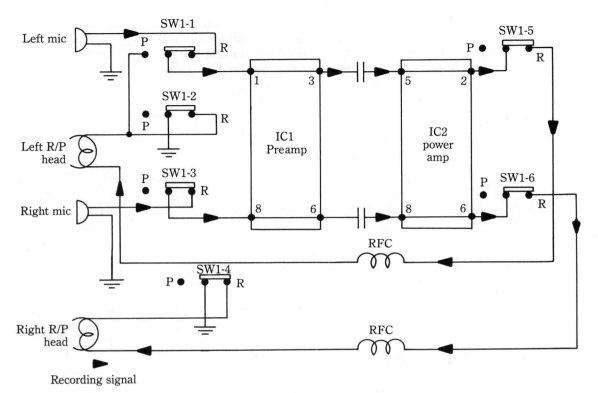

3-4 The dark arrows show the recording signal from the microphone to the R/P heads.

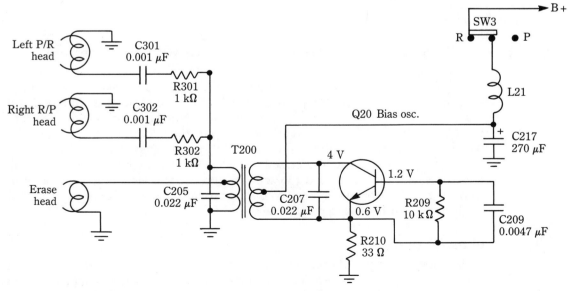

3-5 The bias oscillator excites both the left and right recording heads and the erase head.

When intermittent sound is heard in both speakers, suspect a dual preamp IC, transistors, or a dual output IC. An erratic voltage source applied to the amps can cause intermittent sound in both channels. If your cassette is defective, try another stereo cassette.

After cleaning the R/P heads in both channels, check the audio signal at the volume control for reception of the intermittent channel. If the right channel is intermittent, check the signal at the right channel volume control. Proceed to the audio output amp circuit if the signal is normal at the volume control.

If the audio signal is intermittent at the volume control, clean the control. With no improvement, signal trace the preamp circuits. Check for the intermittent audio signal at the tape head. You will have to crank up the volume on the outside amp or signal tracer because the audio signal is very weak at this point (FIG. 3-6).

Check the signal at the base of the first preamp transistor or IC input terminal. If the signal is normal, proceed to the next preamp transistor or IC. Audio signal tracing on the base, then collector terminal should show an increase in volume. The second preamp transistor has greater audio than the collector terminal of the first preamp. The input and output audio signals of the preamp IC might indicate a noisy or erratic IC. If the audio input signal is normal and the output is intermittent, replace the defective IC (FIG. 3-7).

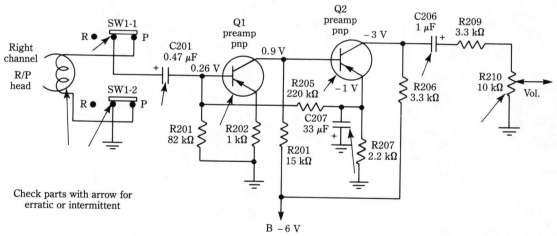

3-6　Check the components that might cause erratic or intermittent problems in the audio channels.

RADIO OR CASSETTE SWITCHING

The audio signal from the stereo circuits are switched into the preamp stages on small cassette players or after the preamp stages (FIG. 3-8). This switching input depends on how many stages of amplification are found in the stereo cassette player. The AM/FM radio switch (S2A) is mounted ahead of the radio/tape switch. Often both switches are small slide switches (FIG. 3-9).

When either AM or FM reception is erratic or dead, spray cleaning fluid into the switch area. Likewise, when either tape or radio reception is intermittent, clean it

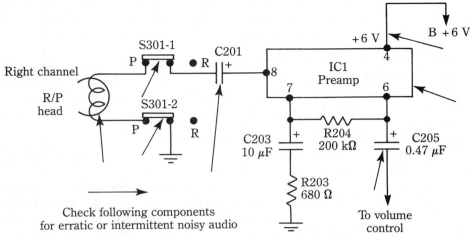

3-7   Check for erratic or intermittent noise in the audio channels.

3-8   Some personal stereo cassette players have FM stereo, such as the Sony Walkman F1.

the same way. Work the switches back and forth to help clean the silver contacts. The stereo FM radio/tape switch might be a DPDT slide switch (FIG. 3-10).

SINGLE STEREO IC

The personal stereo cassette player can have one complete IC for all audio circuits or have a separate power output IC for both channels. Often, the dual volume

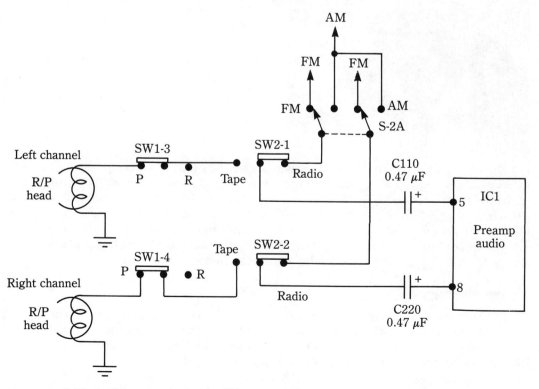

3-9 Clean the AM/FM switch (S2A) or radio/tape switch (SIA) when tape or AM/FM reception is erratic.

3-10 In FM stereo circuits, the FM stereo radio switch might be a DPDT switch.

control is found in the input circuit of the power IC. The audio amp signal is coupled to the volume control with a small electrolytic capacitor. Two separate capacitors couple the output audio signal to the separate speakers or headphones (FIG. 3-11).

3-11 The stereo cassette portable might have one monitor speaker and a stereo earphone jack.

In some players, each headphone switches in and out; the stereo speakers or one large stereo headphone jack does the same. Here, the loud audio signal is cut down with two separate 100-Ω resistors to the earphones (FIG 3-12). Suspect a dirty switch contact when one audio channel is erratic or dead. Clean the earphone switch contact by squirting cleaning fluid into the plug hole. Work the earphone plug in and out to clean the points. The whole stereo jack must be replaced when one side is dead as a result of defective internal switch contacts.

POOR REWIND

The supply spindle or reel is engaged in the rewind mode at a rapid speed. The tape is rotated backwards. Clean the surfaces of spindle drive, idler wheel, and belt drive surfaces. Check for dry spindle and idler wheel bearings. While inside the unit, wipe off all moving surfaces with alcohol and a cloth.

The supply spindle can be fastened to the bearing shaft with a metal, plastic,

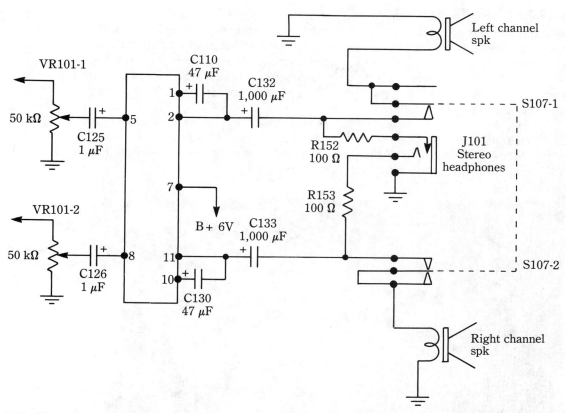

3-12 The dual-stereo earphone jack switches in two 100-Ω resistors for earphone listening when the plug is inserted.

or fiber "C" washer. Remove the small washer and slip off the spindle or supply reel. Clean the reel surface with alcohol and place a drop of light oil on the bearing. Wipe up the excess oil. Replace the supply spindle (FIG. 3-13) if it is broken. If the drive surface spindle is worn or cracked, replace it.

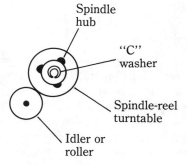

3-13 The supply spindle surface can cause a poor or erratic rewind mode. Clean it with alcohol and cloth.

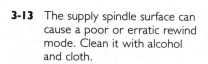

DEAD CASSETTE PLAYER

When none of the cassette player's functions will operate, check and replace the batteries. Insert the ac adaptor to make sure that the batteries or terminals are defective. Doublecheck the polarity of the batteries. It's very easy to insert a battery in backwards. Inspect the battery terminal wires for breaks. Sometimes when the covers are replaced, a battery wire will be pressed against a sharp plastic edge and break into the battery wire.

Next, check the on/off leaf switch. Clean the switch contacts with alcohol and a cleaning stick. Test the switch contacts with the low RX1 range of the ohmmeter. When play is pressed, a direct short should be made across the switch (FIG. 3-14). Inspect the leaf switch for correct alignment.

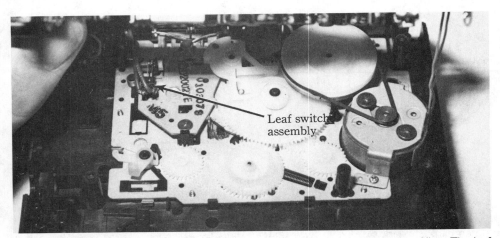

Leaf switch
assembly

3-14   Inspect the on/off leaf switch for dirty contacts or tines that are bent out of line. The leaf switch is under the PC board in Panasonic's RQ-L315.

With no tape action or sound, suspect the batteries or the on/off switch. Rotate the volume control to hear a noise in the speaker. If noise can be heard, the amplifier might be normal with no tape rotation. Check for dc voltage at the small motor terminals. No voltage might indicate a broken wire, dirty motor switch, or regulated motor system.

DEFECTIVE MOTOR

The small motor might be intermittent, erratic, slow, or dead. The intermittent motor might start one time and be dead the next time. Then, when the motor pulley is rotated the motor will begin to run. Erratic motor rotation is often caused by a worn or dirty commutator. These motors are so small to take apart, but sometimes with patience and care, the armature and wire tongs can be cleaned (FIG. 3-15). Other times, you can tap the outside metal belt of motor and it will resume speed.

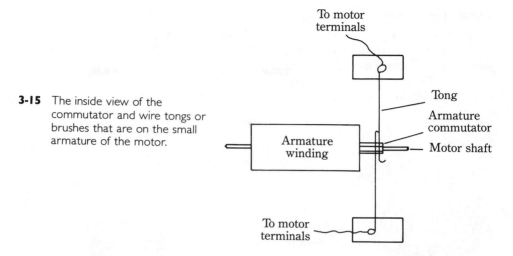

3-15 The inside view of the commutator and wire tongs or brushes that are on the small armature of the motor.

REPLACE DEFECTIVE MOTOR

The small end bearings can be dry and cause slow or erratic rotation. A squirt of light oil in each nylon plastic bearing can help. Be careful not to lose the small carbon brushes or end washers. Some motors can be repaired and others cannot be taken apart. Often, the defective motor is replaced instead of attempting to repair it. The small motor can be held on the main chassis with three end bolts (FIG. 3-16). Replace the motor if a normal dc voltage is applied at the motor terminals and it still doesn't rotate.

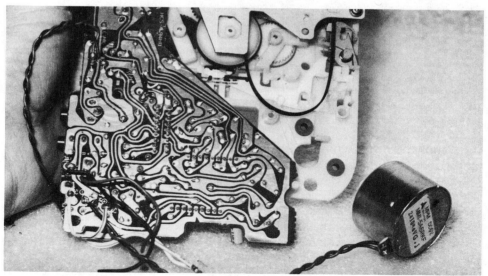

3-16 Three small bolts hold the motor to the metal chassis with the rubber grommet support.

TAPE SPILL OUT

Check for no movement or an erratic movement of the take-up reel when the tape spills out of the cassette as it plays. If the machine is not shut off within a few seconds, the tape can wind around the capstan and pinch roller. Usually, the cassette can be difficult to remove when the tape is wrapped around the capstan. Sometimes the tape must be loose to remove the cassette (FIG. 3-17).

3-17 A Toshiba KTA51 stereo cassette player has two top-side pressure rollers that have auto-reverse action.

Simply play the recorder without a cassette in the holder. Notice if the take-up reel stops or rotates unevenly. If so, clean the rubber drive surface on the bottom of the reel and also clean the drive idler assembly. If the take-up reel is driven with gears or a belt, clean them. Make sure the take-up reel is running smoothly before trying another cassette.

If the take-up reel is rotating normally, suspect a worn or uneven pinch roller. Clean the pinch roller with alcohol and cloth. Sometimes, a sticky substance or too much oxide on the pinch roller can cause tape to pull out. Remove all excess tape that is wound around the pinch roller bearing. Extra tape will make the pinch roller run slow with poor playback (FIG. 3-18).

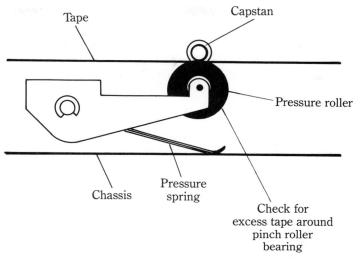

3-18 When tape spills out of the cassette, check around the pressure roller bearings for excess tape. Remove it because the excess tape will slow down the tape action.

Do not overlook a defective cassette when tape spills out of a certain cassette. Try another one and compare them. The tape can be wound loosely or have crimped tape. A squeaky cassette can be caused with dry plastic hub bearings. Insert talcum powder into hub bearing with a toothpick.

NO FAST FORWARD

Fast forward makes the take-up reel spin rapidly at a greater speed than normal play or record. The capstan and pinch rollers are not engaged in fast-forward operation. Clean the fast-forward drive belt, if one is found, with alcohol and cleaning stick.

In most players, fast-forward operation occurs with the idler roller pushed against the take-up reel at a rapid speed. Clean all idler and roller surfaces. Check idler and roller wheels for dry bearings. Slippage between the idler and reel surfaces causes most of the slow or no fast forward problems. Notice if the idler wheel is engaging the bottom roller of the take-up reel assembly (FIG. 3-19). Clean or replace the fast-forward belt that is found in some players.

3-19 When the player has poor or no fast forward, check that the idler surface is pressed against the spindle reel pulley.

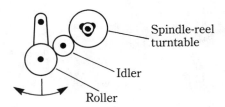

NOISY IC OR TRANSISTOR

Often the noisy transistor or IC can be located in the input and output sound stages. The hissing or frying noise that occurs with low recordings can indicate a noisy solid-state component. Lower the volume and listen for the frying noise. If the noise is present, you know the defective component is between the volume control and the speaker.

Try to isolate the noisy component by grounding the input terminal of the power output IC. Ground the base terminal of a suspected transistor with a 10-Ω resistor (FIG. 3-20). If the noise lowers or disappears, you know the defective component is before this stage. If the noise is still present, replace the noisy IC.

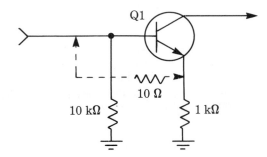

3-20 The noisy transistor can be located by shorting a 10-Ω resistor between the base and emitter terminals.

Sometimes spraying the suspected resistor or IC with cold spray will make the noise louder. Other times, the noise will disappear. In this case, applying heat with a hair dryer on suspected transistor or IC will make the noise reappear after applying cold spray (FIG. 3-21). Do not overlook small ceramic bypass capacitors that can create noise with a B+ voltage on one side. Replace the noisy component with a good part.

3-21 Spray the noisy IC or transistor with a cold spray to make it act up or quiet down, to determine if it is defective.

When the noise disappears with the volume control turned down, the noisy component lies ahead of the volume control. Sometimes grounding the noisy preamp transistor base with a resistor or shorting the emitter resistor can cause the noisy transistor to quit making noise. Start at the preamp input transistor and proceed through the circuit. If the noise is present after grounding out the first preamp signal, the second preamp transistor must be noisy.

Usually, the noisy condition occurs in only one stereo channel. If both channels are noisy, suspect the stereo IC output power chip. The noise might disappear for several days, then reappear. Replace the power output IC when a loud frying or hissing noise is on at all times. A poor internal transistor or IC junction produces most of this noise.

THE PERSONAL PORTABLE CASSETTE PLAYER

Although only a few portable table-top cassette players are manufactured, thousands are still in the field (FIG. 3-22). The cassette player with large pushbuttons was the workhorse of yesterday. Besides regular tape recorder troubles, the large pushbuttons caused many problems. Today's portable cassette recorder has most features including one-touch record, auto-stop, cue and review, tape counter, and built-in condenser microphone. Many of these cassette recorders operated from four C batteries (FIG. 3-23).

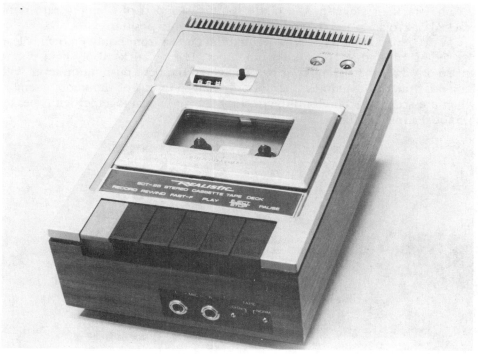

3-22 The table-top portable cassette player was yesterday's workhorse of recording.

3-23   This Panasonic RQ-2103 cassette player operates from four C batteries.

Pushbutton problems

To engage the pushbuttons, extra pressure must be applied. When the switch became stuck or sluggish, the plastic button snapped off. Simply repair the cracked or broken plastic button by cementing it into position.

A dry lever or shifting bar prevented the button from holding down. Clean the sliding bar and lever with alcohol and a cleaning stick. Apply light phono grease or oil on the sliding lever bar. Straighten the bent pushbutton lever with long-nose pliers. Sometimes, the button would pop out of the knob assembly when pressed (FIG. 3-24). Bond the plastic knob and metal together with speaker or model airplane cement.

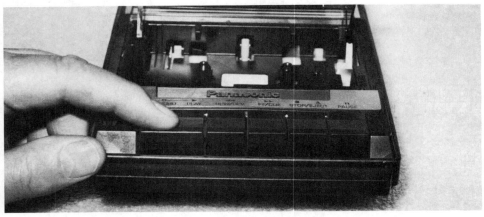

3-24   Cement broken plastic buttons with speaker or airplane glue.

Dirty function switch

Some of these large portable cassette players had a horizontal sliding function switch. Like all sliding switches, the contacts would become dirty and tarnished, making intermittent or no switch contact. Clean them by spraying cleaning fluid in each end of the function switch. Work the switch back and forth by pushing down on the play and stop buttons. A lot of record buttons were broken when the button was pressed without cassette in the holder. More pressure was applied and the button gave in.

Because these switches were fairly large, they had a tendency to cause intermittent or erratic electronic functions. It's best after clean up to solder each switch connection on the PC board. Inspect for cracked or broken wiring around the switch area.

No playback lock in

Suspect a bent or dry lever assembly when the player will not go into any mode or seat properly. If the play button will not seat, inspect lever and catch of sliding bar. The sliding bar is pushed to one side and when the button is seated, the spring pressure applies against the bar and locks the button into the hold position. Check for weak or loss of spring action on the lever bar. Sometimes, the small clip that locks the metal button tap is worn or bent out of line, which releases the button before it is locked into position.

Check for foreign objects in the loading mechanism when any one of the buttons will not function. Like all cassette loading assemblies, the tape heads are pushed forward and locked into position. Clean the loading mechanism and apply light grease on the sliding loading assembly. Excess dust and dirt can fall down into this area, which produces a dry or gummed up mechanism.

No tape action

When the spindles do not rotate and the motor does not sound like it is running, remove the bottom cover. Only four or five small screws hold the bottom plastic cover. Look for a small screw inside the battery compartment. Tape the batteries into the compartment, if needed, and press the play button. Notice if the motor is rotating. Remove the motor belt and give the motor plug a quick spin.

Suspect a dirty or defective on/off leaf switch when the motor and amp are dead. Clean the leaf tines with cleaning solution. Check the switch contacts with the RX1 range on the ohmmeter (FIG. 3-25). Measure the voltage that is applied to the small motor at the PC board. Normal motor voltage indicates a defective motor. These small universal motors are found at electronic and mail-order firms. Check to see if the cost of the motor is more than the recorder is worth before ordering.

Remove the motor terminals from the PC board. Check motor continuity with the ohmmeter. The resistance should be less than 10 Ω. Suspect poor brushes or commutator slip-ring connections if the motor is open. Rotate the motor pulley and notice if the resistance changes. A dirty commutator can cause intermittent or erratic motor rotation. Often, three small screws hold the small motor into position (FIG. 3-26).

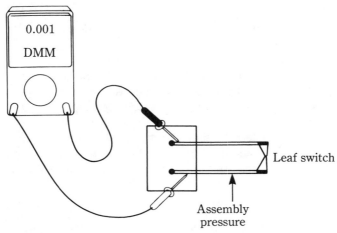

3-25 Check the switch contacts for a direct short in the on position, with the RXI range of the DMM.

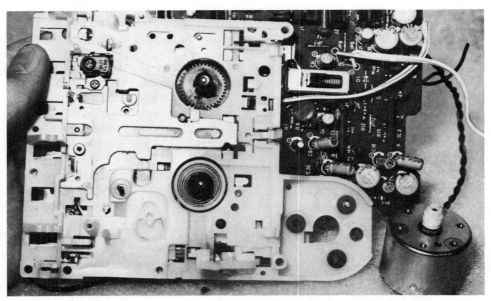

3-26 This defective motor was removed from the chassis by removing three chassis bolts.

Intermittent audio channel

First, clean the stereo tape heads. Clean the earphone and speaker jacks. If you still can't find a sound wave, insert a small screwdriver blade in front of the tape head while in the play mode. When the metal blade passes the tape head, a "thud" should be heard in the speaker with volume wide open. If the tape is rotating, you know power is applied to the motor and amp sections. Eliminate the intermittent speaker by plugging in the earphone.

If you suspect the speaker, clip another speaker across the voice-coil terminals or remove the speaker leads and solder them to another good speaker (FIG. 3-27). Check the volume control for noisy or intermittent sound. Spray cleaning fluid inside the control.

3-27 Disconnect the speaker terminals and clip a good PM speaker to the wires to determine if the speaker is intermittent.

If the player is still intermittent, check the tape head for loose or broken connections. Sometimes, by pressing the pencil eraser against each head terminal, it can show a poor internal connection. Clean the input switch contacts and the external microphone connection for intermittent recording or playback (FIG. 3-28).

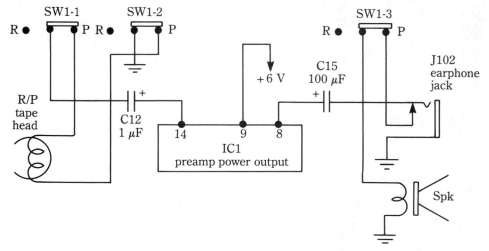

3-28 Clean input switch terminals SW1-1, SW1-2 and SW1-3 for dirty contacts for intermittent recording or playing.

Table 3-1. Cassette Troubleshooting Chart

Problem	Cause	How-to-fix
No sound, no functions	1. Ac power cord	Repair cord
	2. Dead batteries	Check voltage
	3. Bad on/off switch	Check leaf switch for dirty contacts. Clean up.
No ac operation. Batteries okay	1. Defective battery interlock switch	Check switch with ohmmeter
	2. Leaky silicon diodes	Check diodes with DMM diode test.
	3. Bad ac cord	Replace or repair ac cord.
No playback	1. No voltage at amp	Check voltage at leaf switch.
	2. Defective leaf switch	Replace broken switch
	3. Defective transistor	Check transistors for leakage or open.
	4. Defective IC	Check signal in and out of IC.
	5. Bad or open switch	Check and clean contacts.
	6. Defective speaker	Check with ohmmeter for open and replace.
	7. Broken R/P head wires	Replace wires, then solder.
	8. Dirty tape head	Clean with alcohol and cleaning sticks.
No or poor recording	1. Cassette tab broken out	Tape over the opening.
	2. Dirty R/P head	Clean with alcohol.
	3. Defective R/P switch	Clean with cleaning spray.
	4. Defective mic	Replace it.
	5. Defective amp	Check amp voltage and test transistors and IC parts.
	6. Dirty mic jack	Spray cleaning fluid.
	7. No bias voltage	Check bias oscillator circuits.
Poor or no erase	1. Dirty, packed erase head	Clean erase head with alcohol and stick.
	2. Wires off head	Inspect head wires and solder them.
	3. Dirty record/play selector switch	Clean with cleaning spray.
	4. No bias voltage	If dc operated, check the bias voltage.
	5. Defective bias oscillator	Check oscillator transistor and circuits.

Symptom	Possible cause	Remedy
Weak sound	1. Dirty tape head	Clean R/P head.
	2. Amplifier failure	Check supply voltage and amp components.
	3. R/P head switching	Clean switch contacts.
	4. R/P head not resting against tape	Check for missing screw.
	5. Improper azimuth adjustment	Readjust azimuth for maximum output.
	6. Defective tape head	Replace R/P head.
	7. Defective speaker	Check with working speaker.
	8. Mic failure in recording	Substitute another mic.
	9. Excessive or insufficient bias current	Check bias current.
	10. Open speaker coupling capacitor	Shunt with good capacitor.
Poor tone quality	1. Defective transistor or IC	Check with DMM.
	2. Dropped speaker cone	Replace speaker.
	3. Lower dc voltage	Check power supply or batteries.
	4. Poor high frequencies in recording	Make azimuth adjustments.
	5. Dirty tape head	Clean with alcohol.
	6. Poor recording insufficient bias	Check bias oscillator circuits.
	7. Defective mic	Substitute another mic.
	8. Defective cassette	Try another cassette.
	9. Jumbled recording	Clean erase head—check wiring on the head and check the bias current.
Excessive noise	1. Defective transistor	Test the circuit.
	2. Defective IC	Check input and output signals also check voltages on the IC.
	3. Worn volume control	Clean or replace it.
	4. Poor wiring connections	Check the wiring and solder the joints.
	5. Motor rotating	Replace noisy motor.
	6. Hum noise	Replace bad filter capacitors in the power supply.
	7. R/P head magnetized	Demagnetize the tape head.

Inspect the sine wave at the tape head and notice if the sound is erratic or intermittent. If so, proceed to pin 14 of the IC1. When an intermittent is at the input terminal, suspect IC1 or speaker coupling capacitor C15. Monitor the B+ (6 V) applied to pin 9 and notice if the voltage changes or is removed from the circuit. Suspect dirty or poor contacts on the remote switch jack when both motor and amp voltages are intermittent.

Muffled or distorted sound

Often, distorted sound is caused by a dirty tape head, a bad audio output component, or a bad speaker. In the early cassette players, without recording features, the tape head was coupled directly to the audio preamp transistor (FIG. 3-29). No switching is found in the input circuits. The only switching in the push-pull output circuits are the headphone and speaker circuits. Clean the tape head with playback distortion.

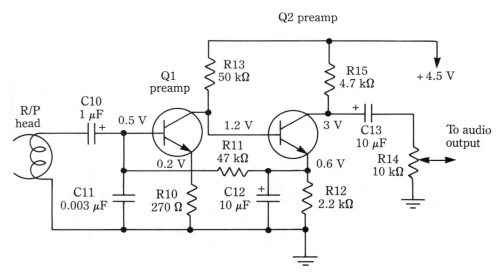

3-29   No switching or recording is in early portable cassette players that have transistor preamp stages.

If the distortion is still present, clip another speaker in the output to determine if the speaker is defective. Next, check the voltages on the output transistor for poor bias or leakage (FIG. 3-30). Test each output transistor for leakage or shorts. Inspect the bias resistors on the base and emitter circuits with the transistors removed from the circuits. Do not overlook shorted or leaky coupling capacitors between preamp and AF driver transistors.

Distortion or muffled sound within the IC1 circuits might be caused by a dirty tape head, a leaky IC, bad coupling capacitors, bad bias resistors, or a bad speaker. Take accurate voltage measurements on every IC terminal. Check the voltage on both sides of C2 and C10 (FIG. 3-31). Replace IC1 if the voltages are

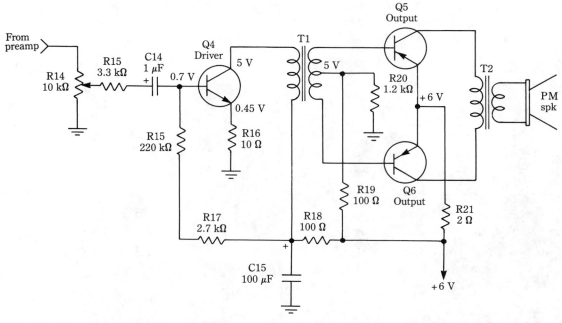

3-30 Distortion is often found in the audio-output stages. Check the transistors and take accurate voltage measurements.

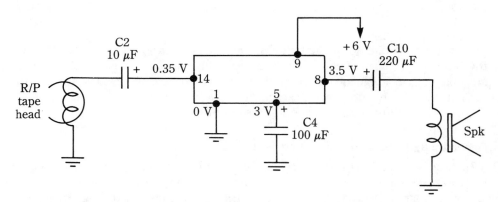

3-31 Check all voltages on the suspected IC and make sure that all components tied to the IC terminals are normal.

fairly normal and speaker has been tested for muffled or distorted sound. Compare the distorted sound within the stereo circuits by comparing the input and output circuits of both stereo channels.

Erratic counter

Suspect jammed gears, a loose belt, or a gummed up tape counter if the counter sometimes works. Often, the tape counter is belt driven from the supply reel. To check it, press the reset to rotate the reel numbers back to zero (FIG. 3-32). Rotate the supply spindle counterclockwise by hand. Notice if the indicator numbers start with number 1 and increase as the supply reel is rotated.

3-32 A counter can become erratic as a result of a worn or loose belt drive from the supply reel.

Suspect a broken or loose belt if no numbers appear on the counter. Remove the bottom cover and the PC board. Inspect the small counter belt. These belts are smaller than a regular rubber band. If the indicator gears are stuck, remove the indicator assembly and spray cleaning fluid into the gears. Slowly rotate the belt drive pulley to ensure that the counter is counting.

Record/play head azimuth adjustment

When the recording is tinny or has muffled audio, suspect a worn, magnetized, or out-of-azimuth adjustment R/P head. Inspect the front of the tape head for excessive worn marks. Degauss or demagnetize the tape head with a demagnetizing cassette. Prepare the unit for azimuth alignment.

Insert a standard recorded 6.3-kHz signal on a test cassette. A 3.5-kHz test cassette will also do the job. Load the earphone jack with an 8-Ω 10-W resistor (FIG. 3-33). Clip the DMM test leads (ac voltage) across the resistor. Adjust the azimuth screw for a maximum reading on the meter. The correct azimuth screw is alongside of the tape head with a small spring underneath one side of tape head (FIG. 3-34). Sometimes this azimuth screw can be reached through a small hole in the plastic cabinet.

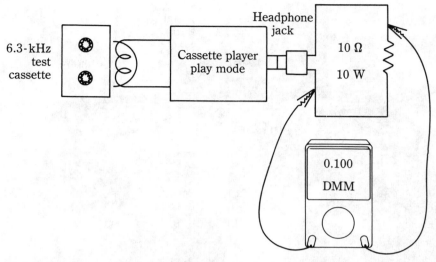

3-33 Adjust azimuth screw while playing a 6.3-kHz test cassette for maximum reading on the ac DMM.

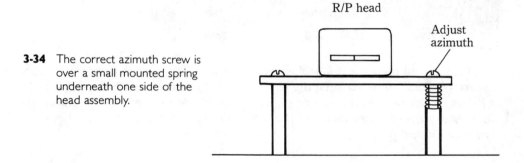

3-34 The correct azimuth screw is over a small mounted spring underneath one side of the head assembly.

Flywheel thrust adjustment

The motor pulley and belt drives a capstan flywheel, which in turn rotates the tape at the pinch roller assembly. Sometimes the end play can become excessive and cause improper speeds. The flywheel can have a small adjustment screw in the metal brace that holds the capstan in place (FIG. 3-35). The metal brace can be bolted to the metal chassis.

Remove the two metal screws alongside the chassis-holding bracket. Check for another screw at other end of the bracket. Remove the capstan thrust plate. Clean capstan bearing and thrust plate with alcohol and a cloth. Remove the capstan/flywheel assembly and clean the capstan bearing hole with alcohol and a cleaning stick. Apply a dash of light grease or phono-lube to the bearing at each end. Not too much, just a small amount on end of cleaning stick.

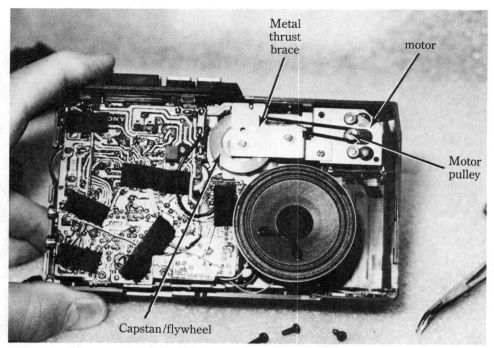

3-35   The capstan/flywheel is held in place with a metal thrust bracket.

Replace the capstan/flywheel and thrust plate. Adjust the screen for a clearance of 0.1 to 0.3 mm between the flywheel bearing and the end plate (FIG. 3-36). If it has no screw, slip a piece of paper between the flywheel bearing and the plate. Adjust the screw or metal bracket against the paper and flywheel. Check the flywheel for end play.

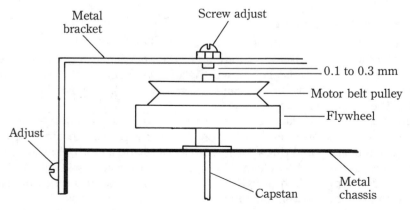

3-36   Adjust the space between the pulley shaft to 0.1 to 0.3 mm.

Chapter 4

Servicing boom-box cassette players

*T*he first large portable AM/FM cassette player contained a cassette player, an AM/FM radio, and a large speaker (FIG. 4-1). After a few years, the AM/FM radio contained a stereo receiver, left and right audio channels, and two stereo speakers. The boom box appeared with high-powered amps and separate speakers. Today, many portable stereo cassette players have detachable speakers and dual cassette players.

The average boom-box portable cassette recorder has an AM/FM stereo tuner, built-in condenser microphone, automatic shutoff, cue and review controls, automatic recording level, high-speed dubbing, and headphone jacks. The boom box can be operated from line voltage or from six C batteries.

The deluxe portable has a digital synthesized tuner with 10 to 20 presets, auto reverse, clock timer, a four- or five-band graphic equalizer, megabass system, dual cassettes, auto recording level, automatic shutoff, a two-way four-speaker system, external microphone input jacks, Dolby B, and up to 10 W output power. The deluxe portable can operate from six C or eight D batteries (FIG. 4-2).

A lot of the latest boom-box players have two different cassette decks. Tape deck 1 is used for recording and playback. This deck can also record from cassette deck 2 or duplicate deck 2. Continuous playback of both decks can be provided with tape deck 2 first and deck 1 last. Each cassette deck can be controlled by its own pushbuttons (FIG. 4-3).

Often, dual tape decks are mounted side by side or one at the top and the other at the bottom. In Sharp's WQ-354, both tape decks are mounted in the cover housing. When the stop/eject button is pressed, the cover assembly moves outward and then down, showing tape deck 2 in front of deck 1. When continuous playing is requested, both decks are loaded with favorite cassettes and the

4-1   A large 5-inch pin-cushion speaker in the Panasonic RQ-542AS AM/FM radio cassette recorder.

4-2   The large deluxe boom box might operate from 12 Vdc or from eight D batteries.

4-3 This GE boom box has dual cassette decks and an AM/FM-MPX stereo amplifier.

door is closed. In this model, long spindles stick out and extend through both cassettes for playback.

CONTINUOUS TAPE PLAYBACK

For extended tape playback of both cassettes, load both cassette holders. Place the tape you want to play first in deck 2. Then, place the second cassette in deck 1. Rotate the function switch to the tape position. Depress the play button of deck 2, then depress the pause button. Now, depress the play button for tape deck 1. Make sure that the dubbing switch is off. When deck 2 has completely played, the automatic stop button in deck 2 engages. The pause button will release and tape deck 1 will begin to play. When deck 1 is completed, the automatic stop shuts off player. Always follow the manufacturers' instructions by the numbers with continuous-play operations.

THREE-BAND GRAPHIC EQUALIZER

The three-slide switch lever control permits you to tailor the bass, mid-range, and treble frequencies to your listening pleasure. More controls are added in players that have the four- and five-band equalizers. Simply moving a control above zero emphasizes the frequency response; moving below zero deemphasizes the frequency response. The low bass frequency range is 100 Hz, mid-range is 1 kHz, and high treble frequency response is 10 kHz (FIG. 4-4). A five-band equalizer can cover the 100-Hz, 330-Hz, 1-kHz, 3.3-kHz, and 10-kHz frequencies.

OSCILLATOR SWITCH

Some large cassette boom boxes or cassette decks have an oscillator switch that is located on the back to eliminate background interference or whistling when

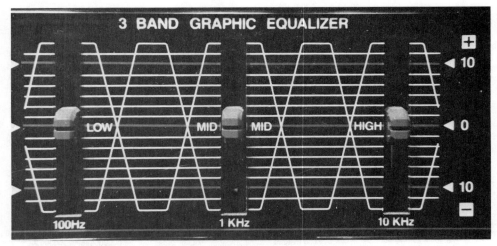

4-4   This 3-band graphic equalizer is found in a GE 2+2 dual boom box.

the radio is turned to the AM band for recording operation. When recording from the AM radio, slide the oscillator switch to either the A or B position to minimize the interference (FIG. 4-5).

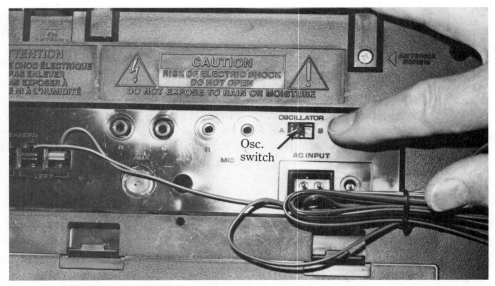

4-5   The oscillator switch eliminates background interference or whistling when recording from the AM band.

VARIABLE MONITOR

A few cassette players, while recording from the radio program, can monitor the radio and have no effect on the recording via the AUX jack. The volume, balance, and graphic equalizer controls can be adjusted to what level you prefer. In some players, the built-in ALC will set the record level for optimum recording level regardless of volume, balance, or equalizer settings.

DUBBING

Some players have a dubbing feature, which enables you to dub (duplicate) a recorded tape cassette in tape deck 2 onto a blank cassette that is placed in tape deck 1. Do not copy or record copyrighted cassettes; that infringes upon copyright laws. Some cassette players have a high-speed, normal, and off switch. With the switch in the high-speed dubbing position, the recording rate is twice the normal tape speed.

Remember, tapes dubbed at high speed will be the same as those recorded at normal speed. Follow the manufacturers' instructions quite closely. Also, the erase head is pivoted out of the tape path in the dubbing mode.

Dubbing from tape deck 2 to tape deck 1

Check for the erase head if you have no instructions or are in doubt of which tape deck does the recording. Place the blank cassette to be recorded in this deck or deck 1. This feature enables you to duplicate a recorded cassette placed in tape deck 2 onto a blank cassette placed in deck 1. Check your instruction book for correct decks because some are marked by numbers and others are marked with letters (FIG. 4-6).

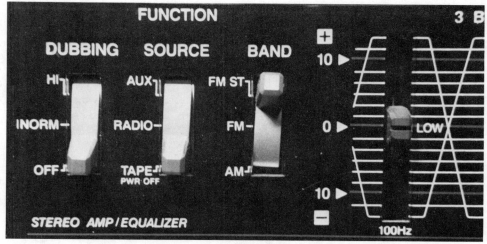

4-6 High-speed dubbing from deck B to deck A with the dubbing switch in the high position.

Place the dubbing switch in the high or normal position. The high position records at twice the normal speed. In some cassette players, dubbing is only provided from deck 1 to deck 2. Place the blank cassette in deck 1. Now, depress the pause button. Then, press the record button.

Drop the recorded cassette in deck 2 and press the play button. When play button in deck 2 is engaged, the pause button on deck 1 disengages and both decks start rotating at the same time. The volume control can be used to monitor the dubbing operation. Do not change the dubbing switch position.

When you have finished the dubbing operation, press deck 1 and deck 2 stop/eject buttons. If the whole cassette is to be duplicated, let the automatic stop of each deck shut off the player. After the dubbing operation has been completed, place dub switch in the off position.

THE ERASE HEAD

The erase head function is to remove any previous recording before the tape moves across the record/play tape head. Jumbled or excessive distorted music can be caused with packed dust oxide on openings of tape head, which prevents erasing the previous recording. Usually, only one erase head is contained in the cassette compartment. This head is smaller than the record/play head and is located to one side. The R/P head is mounted in the center.

In dual-cassette system, the recording deck contains the erase head. The erase head can be bolted next to the R/P head. Within Emerson's CTR961, the A and B dual-deck system, the erase head is found in the bottom B deck. Tape deck A is mounted at the top. The erase head in the bottom deck operates on a pivot and stays down out of the way while dubbing and playing. When the record button is pushed, the erase head is moved upward to engage the tape (FIG. 4-7).

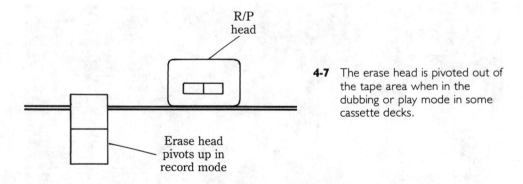

4-7 The erase head is pivoted out of the tape area when in the dubbing or play mode in some cassette decks.

Most erase heads are excited with a dc voltage or with an oscillator bias circuit. The dc voltage is switched in record mode to the erase head (FIG. 4-8). This voltage comes from the dc power supply or batteries.

When the record button is pressed in the deluxe boom-box player, B+ voltage is applied to the bias oscillator. This bias signal is applied at all times to both stereo R/P tape heads and can be seen with an oscilloscope (FIG. 4-9). Poor recording or no recording can be noticed when the bias signal is removed from the erase and stereo R/P tape heads. A defective bias oscillator circuit (FIG. 4-10) can jumble or distort music while recording.

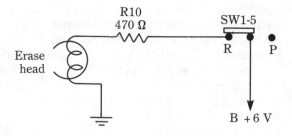

4-8 The dc voltage is switched to the erase head when recording.

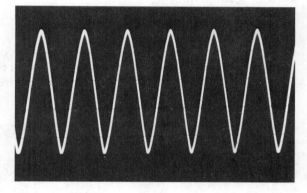

4-9 The bias waveform at the high side of the R/P tape head.

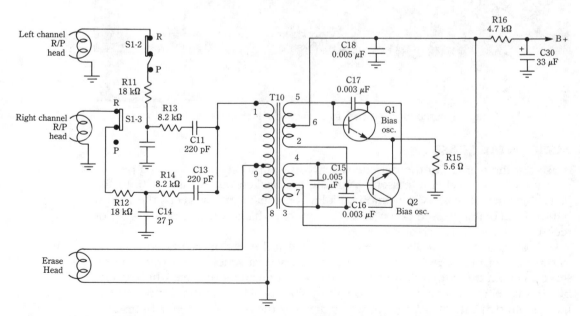

4-10 Check the bias oscillator circuits when you find muffled, distorted, or jumbled recording.

The larger expensive boom box can have an oscillator coil with a dual primary winding. The erase head is excited from the center top while both left and right stereo recording heads are excited from the top winding through a resistor/capacitor network. Q1 and Q2 are dual oscillators with B+ applied in record position.

Most bias oscillator problems are caused by a leaky or open transistor, dirty record/play switch, or poor transformer board connections. First, clean the erase head with alcohol and a cleaning stick. If the erase head does not remove the previous recording, check the voltage that is applied to Q1 and Q2. Test transistors Q1 and Q2 in the circuit. Check the continuity of each transformer winding. If one is open, solder the transformer board connection. Replace the transformer if a winding is open.

ERRATIC SWITCHING

When the boom box is operated and fails to record, suspect dirty record/play switch contacts. Clean all of the switching contacts on the entire switch assembly with cleaning spray (FIG. 4-11). Place the nozzle inside by each contact and spray all switching wafer contacts. Work the switch back and forth to clean the contacts.

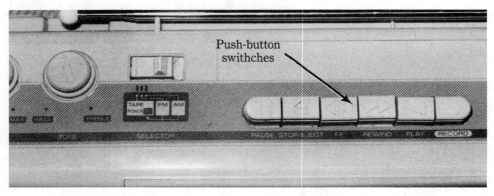

4-11   Clean the pushbutton switches to correct operation and erratic switching.

ACCIDENTAL ERASE

Break out the rear tab of the cassette when you want to keep a recording permanently (FIG. 4-12). When a cassette with the tab broken out is inserted into the player, accidental erase is prevented by a lever mechanism that keeps the record button from being pressed down. If the symptom is that a certain cassette will not record, first check the tab at the back.

Most recorded musical cassettes have no tab at the rear of the cassette. This is to prevent erasing or recording over the purchased cassette. If you want to preserve a certain cassette, remove tab at the back. You can record over this cassette by placing cellophane tape over the opening. The small lever at the rear of cassette holder can be pushed and held in while pressing the record and play button for proper operation. Just insert another cassette with tab to make sure player will record.

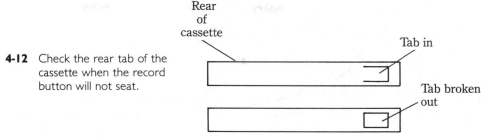

4-12 Check the rear tab of the cassette when the record button will not seat.

BROKEN SOFT EJECT

In some models, the cassette lid will slowly open when the eject button is pressed. The soft-eject system can operate from a gear type or plunger arrangement. In newer models, the cassette lid slowly opens with a lever-gear arrangement (FIG. 4-13).

4-13 A soft-eject mechanism might consist of a braking gear and plastic gear assembly fastened to the cassette door.

Suspect a broken gear or suspect that the gear lever is off track if the door pops open with soft-eject system. The gear lever is attached to the cassette lid and rotates a small plastic gear that has braking action. If the door is difficult to open, suspect a gummed or damaged gear assembly. Clean the gear assembly for smooth ejection with alcohol and a cleaning stick. Do not apply light oil to the braking gear assembly (FIG. 4-14). Remove any thread, twine, or foreign material from gear assembly if it will not open.

NO ACTION ON DECK 2/DECK I OKAY

When one deck is operating and the other will not rotate, tape suspected defective motor or broken motor belt. Each tape deck has its own motor and drive system (FIG. 4-15). Measure the voltage that is applied to the motor terminals. Rotate

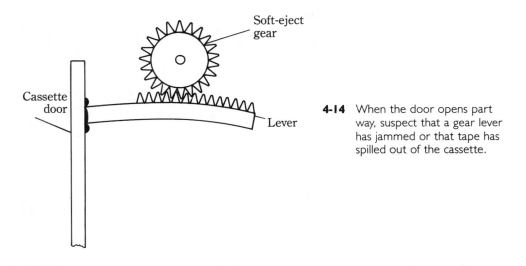

4-14 When the door opens part way, suspect that a gear lever has jammed or that tape has spilled out of the cassette.

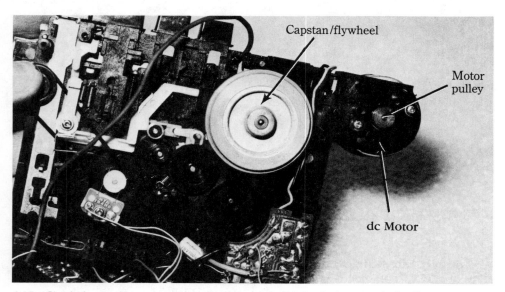

4-15   Check the motor, motor drive belt, and capstan/flywheel when one deck will not move the tape.

the belt if voltage is present and see if the motor starts to rotate. Replace the motor if it will not rotate with voltage applied.

Often, when the play button is pressed, you can hear the small motor rotate by placing ear next to the cassette door. If the motor is running with no tape motion, suspect a broken belt (FIG. 4-16). Notice if the capstan is rotating. Place your finger over the capstan to see if it's turning without the pinch roller engaging. Notice if the pinch roller is touching the tape.

4-16 Each tape deck has its own motor, capstan/flywheel, and pinch roller.

Sometimes, the capstan/flywheel bearing will run dry and freeze. Simply move the flywheel by hand. If it starts off slow, suspect dry bearings. Remove capstan/flywheel assembly and mounting bracket. Clean the bearings and lubricate them with light grease. Check the motor belt for broken, loose, or stretched areas. Replace the motor belt if it shows signs of slipping.

POWER SUPPLY CIRCUITS

If no voltage or if you find an improper voltage at the cassette motor and amplifier, suspect a defective power supply. Does the unit play on batteries? If so, go directly to the ac power supply. Very seldom will you find any fuses in these power supplies. The early power supplies might have full-wave rectification with only two diodes (FIG. 4-17).

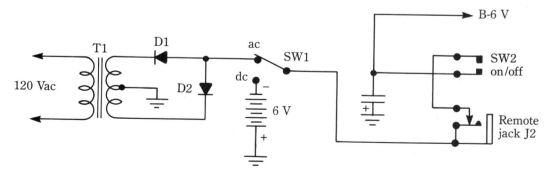

4-17 The early cassette player might have full-wave rectification with two silicon diodes.

The low-priced power supply might consist of only two silicon diodes, a transformer, and an electrolytic capacitor. The power-supply output can be switched into the circuit with a remote switch and a leaf switch. Check the silicon diodes with the diode test of the DMM. Sometimes when one or more diodes are shorted, the primary winding of the power transformer (FIG. 4-17) is taken out. Because the secondary windings are wound with heavier wire, this winding rarely opens. Check both windings with the RX1 ohmmeter range.

You might find the two small diodes inside one plastic component, such as the bridge rectifier. In other supplies, separate silicon diodes are used. Replace the entire part when one is shorted or open.

Switch the player to ac operation and measure the dc voltage across the electrolytic capacitor. If the voltage is low, suspect a leaky component, such as an output transistor or IC. Low voltage can be caused by a weak or open filter capacitor (C1). Shunt another 2,000-μF 25-V electrolytic capacitor across C1. Clip it in with alligator leads with the power cord pulled from the outlet. Replace the defective electrolytic capacitor if the voltage is restored. Observe correct polarity (+) of the capacitor when soldering it in. Check SW1 and the remote jack for poor contacts.

BRIDGE RECTIFIER CIRCUITS

In bridge rectifier circuits, four diodes provide full-wave rectification. Usually the transformer has only a primary and secondary winding without a center tap. The low ac voltage is applied across the bridge circuit and the dc voltage is taken from the two positive ends of diodes that have been tied together (FIG. 4-18). The bridge diodes can all be found in one unit or separately. When one or more diodes are defective, the component(s) must be replaced (FIG. 4-19).

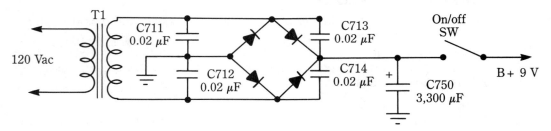

4-18   The bridge-rectifier circuit is used in the power supply of many cassette recorders.

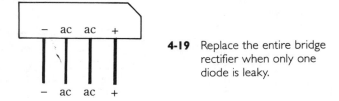

4-19   Replace the entire bridge rectifier when only one diode is leaky.

When the ac cord is inserted into the outlet socket, the ac/dc switch switches the batteries out of the circuit. Now, an ac voltage is applied to the primary winding of the power transformer. An external dc socket is in series with the positive lead of battery (FIG. 4-20).

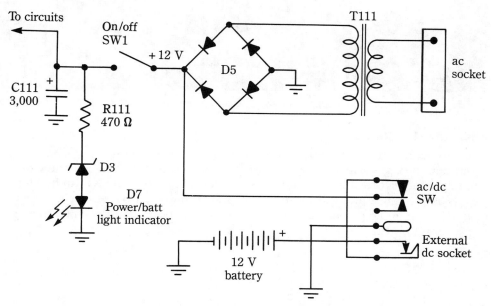

4-20 An external dc battery source or supply might be used with the external dc jack in the low-voltage circuits.

A poor contact in either the ac/dc switch or in the external dc socket can prevent battery operation. Sometimes when the unit is dropped on the ac cord outlet, the plastic socket splits and can misalign a leaf switch. Sometimes, these switch assemblies can be removed, repaired, and placed back together with epoxy cement. The original part might be difficult to obtain.

If the cassette player has a power/battery light indicator, you can tell if the power supply or battery source is normal with the lighted LED. If nothing occurs in battery operation, the light might be dim with one or two defective batteries. Likewise, when switched to ac operation, an unlit power light indicates problems in the low-voltage circuits.

Take a quick voltage test across the electrolytic capacitor (FIG. 4-21). Suspect a defective off/on switch, bridge diode, transformer, or ac cord if no voltage is available in ac operation. Always check the battery operation to determine if the ac power supply is defective. Suspect the external dc socket switch, ac/dc switch, or dirty or corroded battery terminals when batteries test normal. Make sure that the ac cord is removed for battery operation on some models.

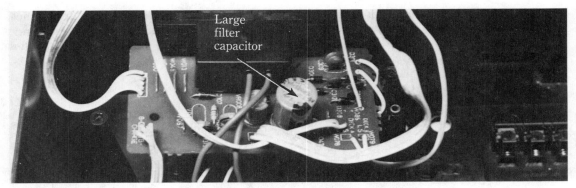

4-21   Check the dc power supply with a voltage measurement across the large filter-capacitor terminals.

RECORDING CIRCUITS

The recording circuits in early boom boxes had transistors in the preamp and audio output stages. The preamp transistor circuits consisted of two or three preamps and one record amp transistor. The condenser microphone is switched into the preamp circuits and the picked up sound is amplified by several transistor channels. The audio preamp signal is switched to the base circuit of a record amp transistor (FIG. 4-22).

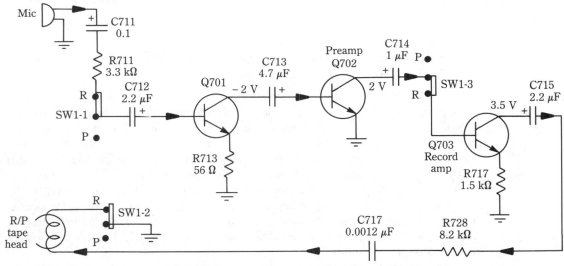

4-22   A block diagram of the recording circuits switched into the transistorized cassette circuits.

The recorded output is capacity coupled through several resistor/capacitor networks. The recorded signal is applied to the R/P head and ground. At the same time, the R/P head is excited with bias voltage from a one-transistor bias oscillator. Also, the erase head is switched into the circuit in record mode. The erase head can be bias operated or operated from the B+ source (FIG. 4-23).

Clean the R/P head when the machine will not record. Clean the R/P switches (SW1-1, SW1-2 and SW1-3) with cleaning fluid. If the recorder operates normally when playing, suspect the microphone or Q703. Both Q701 and Q702 are in the circuit in playback mode. The only transistor not included in playback is the recording amp (Q703). Any malfunctioning part in the recording path from SW1 to SW3 to the R/P tape head can prevent recording (FIG. 4-24).

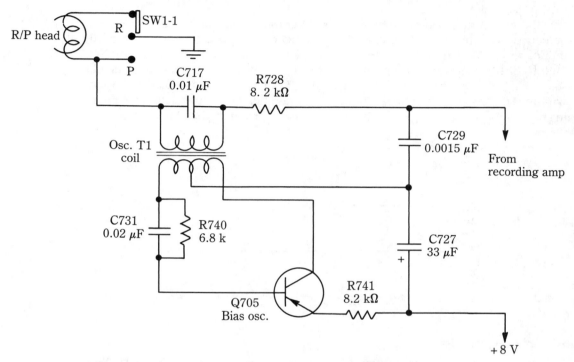

4-23 A simple bias-oscillator circuit that excites the R/P tape and erase heads.

4-24 An electronic technician checking the recording circuits of a boom box.

Check Q703 in the circuit with the transistor tester of the DMM. Take voltage measurements on all terminals of Q703. Shunt C715 and C717 with respective capacitors. Remember, the microphone pick-up audio can be signal traced through the preamp and record stages up to the R/P tape head with the external amplifier. Place a radio speaker near the mic and signal trace each stage with the audio amp and a pair of headphones. A defective bias oscillator stage can prevent a quality recording.

If the recording sounds jumbled with other recordings on the tape, suspect a defective erase head. The dc erase head is switched into the circuit in record mode, with SW1-5 (FIG. 4-25). B+ 8 V is applied to the tape head through R787 (330 Ω) of the erase head and ground. If the voltage is low across the erase head, suspect a poor record-switch contact at SW1-5. Check the erase head for open windings or broken head wires. Notice if the erase head is riding against the passing tape. A loose mounting or a missing screw can prevent the head from touching the tape.

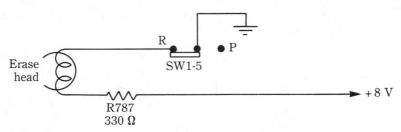

4-25   The dc-operated erase head is in the recording circuit when it is switched to record.

IC recording circuits

You can find ICs in the preamp and in both left and right stereo channels. The amplified audio signal from the microphone and preamp IC can be fed back to the corresponding R/P head in record mode after the preamp stage. Of course, the erase head and bias oscillator are switched into the circuit when recording (FIG. 4-26).

The two stereo microphones can be fed through an external mic jack and switched to the preamp, IC1101. The left-channel audio signal is switched into pins 9 and 10 of the preamp IC. The right-channel mic audio signal is switched into pins 6 and 5 of the preamp. Notice that the preamp stages serve both record and playback modes.

The output signal from the left channel at pin 13 is coupled through a capacitor (C205) and switched to the left R/P head. The output signal of the right channel at pin 2 is fed to C305 and switched to the right R/P tape head. When recording, the dc-operated erase head is switched into the record circuit with SW1-7. dc bias is applied to each left and right channel R/P head through R210 and R320 (FIG. 4-27).

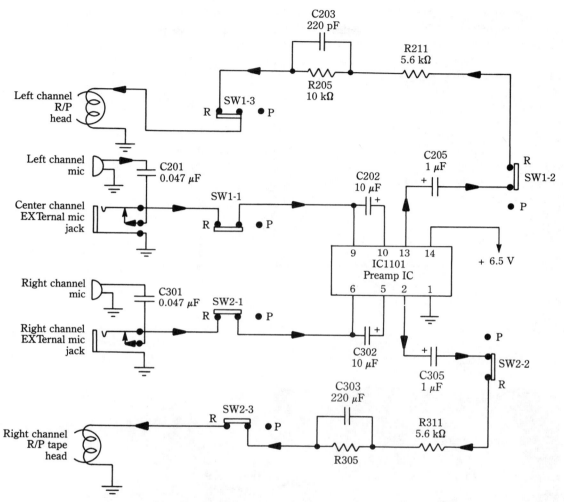

4-26 The IC recording signal path in the record mode.

TROUBLESHOOTING IC RECORD CIRCUITS

Before attempting to repair the recording IC circuits, determine if the playback circuits are normal. If the playback circuits are normal, clean the record and erase heads. Clean any switches that are related to the recording circuits. Check the oscillator bias circuits when you notice distortion or weak recording. Measure the dc voltage or bias voltage at the erase head when music is jumbled (FIG. 4-28).

Check the schematic for another record IC component. Larger boom boxes have a separate recording IC after the preamp circuits. Measure the supply voltage (V_{CC}) from the low-voltage power supply. If the voltage is very low, the recording IC might be leaky or the supply voltage might be improper. Remove the supply

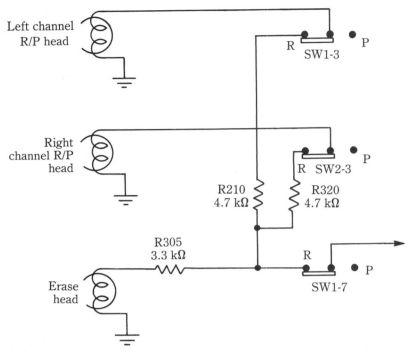

4-27   The dc voltage is switched to the R/P and erase head when it is excited with a dc voltage.

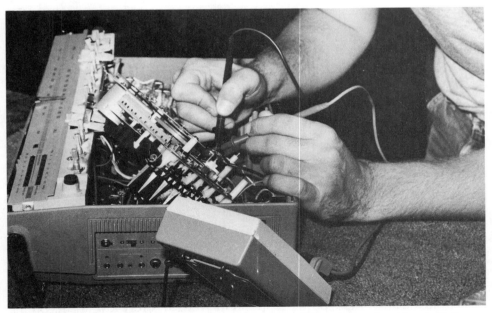

4-28   An electronic technician checking dc voltages on an R/P and erase head in a J. C. Penney 681-3950 boom box.

voltage from the IC by unsoldering the IC pin from the PC wiring. If the voltage rises above or near to the required voltage, the recording IC is leaky, so replace it.

Check all voltages on each terminal of the IC and compare them to the schematic. If one or two pin voltages are way off, check the components that are tied to these pins for a possible shorted capacitor or a change of resistance. If the recorded signal is coming into the IC and not at the output terminal, replace the leaky IC.

Signal trace the audio signal with an external amp or with a signal tracer from the microphone to the recording amp IC. Place a radio or another cassette player (with a tape playing) in front of the microphone. Check the signal at the input and at the output pin of the suspected IC. Replace the IC if the voltage and input signal is normal, but no audio output from the IC.

DEAD RIGHT SPEAKER

The voice coil of a boom-box speaker can be damaged with excessive volume or shorted audio output circuits. The defective speaker might be dead, mushy, intermittent, or distorted. Clip another speaker across the terminals of the dead speaker. If another PM speaker is not handy, interchange the leads that go to the normal speaker for test only (FIG. 4-29).

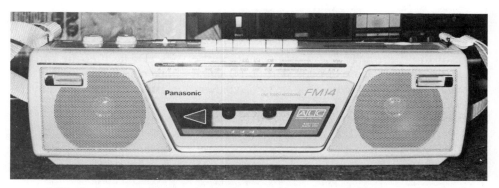

4-29 Exchange speaker terminal leads to test the suspected dead stereo speaker.

Suspect a damaged voice coil with a dead, intermittent, or noisy speaker sound. The intermittent speaker can be located by applying equal pressure on the speaker cone with music playing. Pushing the speaker cone in and out makes the sound cut up and down. Check for an open voice coil with an ohmmeter at RX1 range. Check speaker hinge connections on some speakers that hang alongside and use the hinges as speaker connectors.

Large speakers can be damaged with constant excessive volume and blown voice coils. Push up and down on the center of the speaker and notice if the voice coil is dragging. The muffled sound can be caused by a frozen voice coil on the center magnet. The voice coil will not move when it is frozen to the magnet. When the dc voltage is applied to the voice coil in a dc-coupled speaker circuit, the voice coil can be burned open (FIG. 4-30).

4-30 A dc voltage or excessive volume applied to the voice coil can even damage high-wattage speakers.

Small holes punched into the speaker cone can be repaired with speaker cement. If the speaker is "blatting" on bass notes, the voice-coil diaphragm support or speaker cone can come unglued. Apply speaker quick-drying cement under the cone area. Replace any speaker that has large holes, a torn cone, or a damaged voice coil.

DEFECTIVE RECORDING METERS

In some boom boxes, a separate recording meter or LED circuit determines the range of the recording. The meter circuit can be switched in after the preamp stage or after the AF amp circuits. The audio signal can be checked right up to the diodes of the meter circuit. One or two separate transistors can be found in the meter circuits (FIG. 4-31).

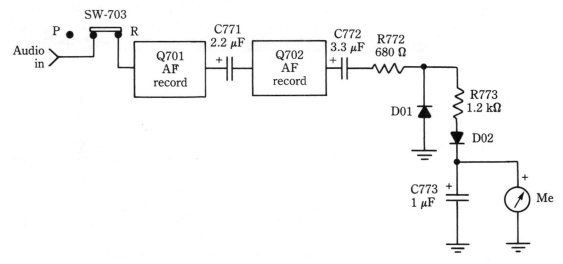

4-31 A block diagram of a recording meter in the output of a rectification circuit.

When the recording meter is not moving in when recording, check the audio signal through the AF stages to the small meter diodes. Trace the audio signal with separate external amp or with a signal tracer. Check the meter for an open winding if the audio is traced up to the meter circuit. Test each diode with the diode test of the DMM. The meter hand might be bent or rub against the dial area and not move with the recording.

LED meter circuits

The peak-level LED display meter can be driven with an IC. The audio signal is applied to the input terminal with LEDs in the output audio-level circuits as indicators (FIG. 4-32). As the recorded or audio level increases, the LEDs will light up accordingly. You can find one IC for each set of display LEDs in stereo channels.

Signal trace the audio signal up to the input terminal of IC501. If audio is found at input terminal and no LEDs light, suspect improper supply voltage or a defective IC. Measure the dc voltage (12 V) going to one side of all LEDs. Suspect an IC if the supply voltage is normal. Check the voltage that is applied to each LED.

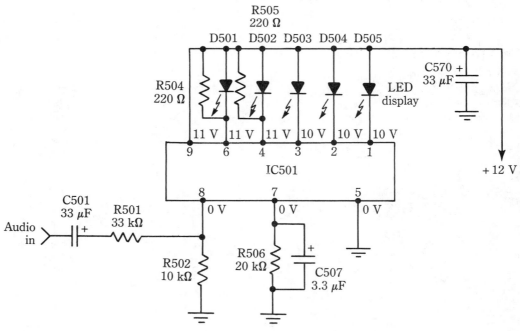

4-32 An LED meter-level display is operated from one large IC.

The LED array might have all LEDs in one component. In earlier units, separate LEDs were mounted. If one signal LED will not light with correct applied voltage, replace it. If one LED will not light in the LED array, replace the entire component (FIG. 4-33). Each LED can be checked with the diode test of the DMM. In stereo circuits, check the defective LED indicator circuit with the normal channel.

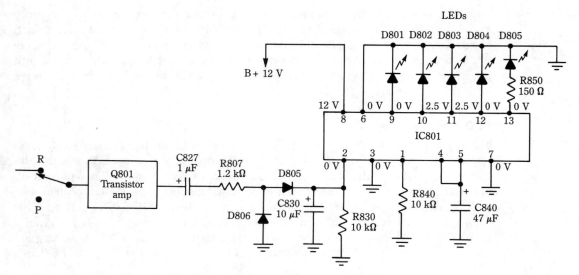

4-33 You might find single LEDs in a recording or sound-level meter circuit.

Chapter **5**

Troubleshooting portable AM/FM cassette/CD players

Besides the AM/FM stereo tuner, the cassette player can have a compact disc (CD) player. The CD player can be located at the top or front of the portable cassette system. Dual cassette decks are featured in some models. Auto reverse, remote control, Dolby B, and clock/timer operations are found in the deluxe units.

The programmable CD player can have up to 36 tracks. Detachable speakers and 10 watts of power provide quality sound. The auto-search music system restarts a CD selection or plays the next one at the touch of a button. Most of the portable cassette-CD players require 8 or 10 D batteries and 4 AA batteries.

THE POWER SUPPLY

Besides operating from batteries, the player can be operated from a built-in power supply. Most players use a full-wave bridge rectifier system with a fairly large filter capacitor (FIG. 5-1). Each silicon diode is bridged with a 0.022-μF bypass capacitor to eliminate diode interference. D401 prevents accidental polarity that is played into the external dc jack (J402). C325 provides adequate dc filtering with on/off switch S305 in series with the power supply, amplifier, and motor.

The cassette motor operates directly from the 9-V battery or power-supply source. Isolating choke coil L301 and capacitor C326 provide extra filtering for the small cassette motor. C328 (0.1 μF) provides bypass arcing of the small commutator to ground (FIG. 5-2). Some expensive chassis have motor transistor regulation or voltage-regulator circuits.

If on/off switch S305 is dirty, the cassette player might be erratic or dead. Try all tape, CD, and radio functions before digging into the chassis. If the radio and CD player operates but the cassette player doesn't, suspect that the function

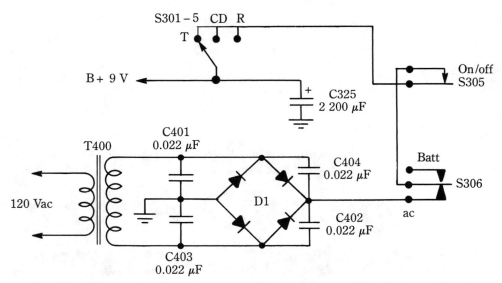

5-1   The full-wave bridge ac power supply with on/off switches S305, S306, and S301-5.

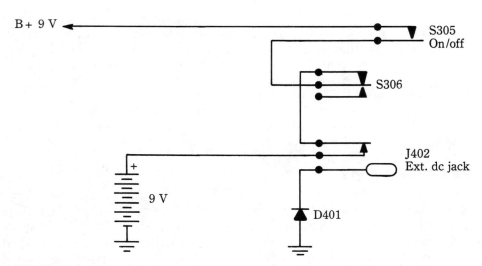

5-2   The battery is switched out of the circuit when external jacks J402 and S306 are used.

switch is dirty. When the unit operates on batteries and not ac power, suspect a defective ac cord, dirty S306 switch contacts or dirty power-supply circuits.

Check the dc voltage across the large filter capacitor C325 (2,200 μF). Very low dc voltage might indicate overload leakage from the radio, tape, or CD circuits. Switch to all circuits to see what section is pulling down the voltage. Then, isolate the overloaded circuits. If the voltage is low on all sections of function switch, suspect a poor power supply.

Test each diode with the diode test of the DMM. If one or two diodes are leaky, replace the whole bridge component. With normal diodes and no dc output voltage, suspect an open primary winding of T400. Low dc output voltage can be caused by a leaky or open filter capacitor (C325).

Measure the resistance across the C325 terminals. Suspect a leaky capacitor if the measurement is below 250 Ω with S301 switched to the radio function. Clip another 2,200- or 3,300-μF electrolytic capacitor across C325 with the power off to determine if the capacitor is open. A hum in the music when the volume is turned down indicates a dried-up filter capacitor C325.

HEAD CLEAN UP

Open the cassette door and hit play. This will push out the R/P and the erase head for easier cleaning from the door opening. In some models, two etcheon screws at the front of cassette door let the front plastic loose for easier tape head cleaning.

Clean the tape head with a cleaning stick and alcohol. Difficult tape oxide that is packed on the front of the tape head can be removed with a wood dowel or a pencil eraser. Make sure that all small gaps on the front of the tape have a clean surface and are not clogged with tape residue.

At the same time, clean the pinch roller and capstan. Sometimes, when a pulled out tape is wrapped around the capstan, the packed oxide dust can be difficult to remove. Make sure that all tape oxide is removed from the pinch roller. Hold the roller in one spot, clean it, then move to another section of the rubber roller. Do not apply so much alcohol on the cleaning stick that it runs down inside the capstan bearing (FIG. 5-3).

5-3 A close-up view of the pinch roller in the cassette player. Clean the pinch roller after cleaning the tape heads.

ERRATIC PLAY

Erratic play or fast-forward modes can be caused with improper torque, a dirty or binding flywheel, dirty or loose drive belt or clutch assembly. Clean the flywheel, motor pulley, drive belt, and clutch assembly with alcohol and a soft cloth. The capstan and fast-forward/rewind belts should not be cleaned with alcohol and cloth. Replace belts that have been stained with grease or oil.

Fast-forward torque adjustment

Insert a torque-adjustment cassette in the holder (FIG. 5-4). Measure the fast-forward torque. A torque of more than 50 g/cm is necessary for fast-forward operation. Check the manufacturer's service manual for correct torque adjustment. If the fast-forward measurement is under 50 g/cm, wipe the flywheel, clutch assembly, and/or replace the drive belt.

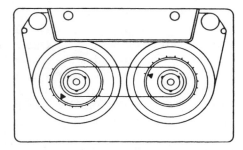

5-4 Check the fast-forward torque with a torque cassette. A torque of more than 50 g/cm is necessary for fast-forward operation.

Take-up torque adjustment

Insert a cassette torque meter and measure while playing. If the take-up is not adequate (30 to 60 g/cm), wipe the flywheel, belt, and take-up reel assembly. Replace drive belt if it is cracked or loose. Poor take-up torque might cause the tape to spill out of the cassette.

Rewind torque adjustment

Insert cassette torque meter and if torque is below 50 g/cm, suspect dirty or worn parts. If the rewind torque is not over 50 g/cm, wipe the flywheel, clutch assembly, and/or replace the drive belt. Low rewind torque can cause slow or erratic rewind speeds.

PINCH-ROLLER ADJUSTMENT

After cleaning the pinch roller, it's wise to check the pinch-roller adjustment. In the playback mode, measure the pinch roller contact with a spring gauge (0- to 500-g gauge). Hook the spring gauge to the pinch roller and pull it away from the capstan (FIG. 5-5). Measure the force at the moment when the pinch roller comes in contact with the capstan (when the pinch roller starts revolving). The gauge should read from 100 to 320 g. To adjust the contact pressures, bend the spring (27) and/or replace the spring. Remember, each manufacturer can have its own specifications, but most will read within the above measurements.

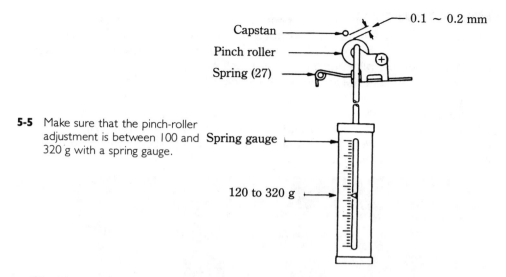

5-5 Make sure that the pinch-roller adjustment is between 100 and 320 g with a spring gauge.

NO FAST FORWARD

Does the tape move normally in the play mode? If the tape is slow in both fast forward and play, clean the capstan, motor belt, and motor pulley. Suspect a defective motor or binding capstan if the tape is still slow after clean up. Remove the motor belt and give the flywheel a spin with your fingers. A sluggish or dry capstan will not rotate a complete turn. Remove the capstan/flywheel, clean the bearings, and lubricate the bearings with light oil or grease.

Suspect an oily or slick fast-forward drive belt when the playback is normal and the fast forward is erratic. Some units have both a motor drive belt that rotates the capstan and a separate belt that drives the fast forward (FIG. 5-6). Clean the fast-forward drive belt and all drive surfaces with alcohol.

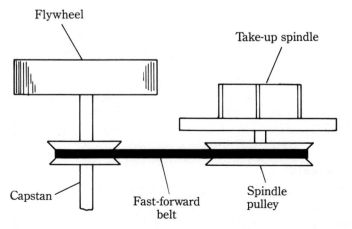

5-6 The fast-forward spindle might be operated with a drive belt between the fly-wheel pulley and the bottom of the take-up reel.

BINDING BUTTONS

After several years of usage, the pushbutton assembly might become dry and sluggish, which results in some buttons not seating properly. Suspect a dry loading assembly when the play button does not lock in. Clean the levers and the surface sliding parts with alcohol and a cleaning stick. Check each button lever for dry or binding parts.

Sometimes spraying cleaning fluid into the button assembly can help. Be careful not to spray so much liquid into the assembly that it drips onto belts and driving surfaces. Often, when one button assembly sticks, the plastic button breaks off. Cement the button back in lever assembly (FIG. 5-7).

5-7 Squirt cleaning fluid between the buttons that are binding against the plastic.

NO TAKE-UP REEL ACTION

The sluggish or erratic take-up reel results in pulling or spilling out tape in the capstan and pinch roller area. The take-up reel must rotate smoothly to take off the tape from the capstan to load the take-up spindle. If the spindle or reel stops or operates in an erratic or intermittent motion, tape can spill out and destroy a good cassette.

Hit play and watch the rotation of take-up reel (on the right). Clean the idler, clutch assembly, drive belt, and capstan assembly. On units that have a gear-train drive spindle assembly, clean the gears and dry bearings. Place a drop of light oil on each gear bearing. Often, poor or no take-up reel action results in no fast forward in lower priced cassette players.

REMOVING COVERS

Although it might appear to be difficult to remove the back cover of the portable CD-cassette player, only a few screws hold the cover to the front cabinet. Always remove the battery cover and all batteries. Remove the several screws that hold the rear cabinet (FIG. 5-8). Remove the five screws that hold the power-supply board.

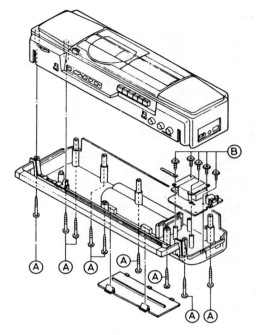

5-8 Remove screws (A) to remove the back cover on the CD/cassette player.

To disassemble the whole cabinet, remove the two screws that hold the PC board. Remove the three knobs (balance, tone and volume) from the front. To remove the cassette player assembly from the cabinet, remove the screws that hold the large amplifier PC board. Now, remove the screws that hold the cassette mechanism (FIG. 5-9).

Remove the screws that hold the CD mechanism. Remove the three screws that hold the CD support and gear holder. Finally, remove the three screws that hold the display PC board. Lay out all the parts in a row so that they can be replaced properly. Sometimes listing the various components on a piece of paper can help to reassemble the mechanism. Check FIG. 5-10 on how to remove the speakers and the cassette door assembly.

BLOCK DIAGRAM

Always check the block diagram before attempting to service the defective player. Try to isolate the defective component from the block diagram (FIG. 5-11). For instance, in this Radio Shack 14-527 compact disc and cassette player, a separate IC (IC301) is used in record mode. If the cassette player played perfectly, but would not record, IC301 might be the defective part. In most cassette players, the preamp and AF amp stages are used for both record and play.

Notice that IC302 is a Dolby circuit IC. Here the record and playback signals go through IC302. It's possible that the stereo playback and record signals enter IC302 at pins 1 and 14. If the cassette plays, but doesn't record, check the record signal out of pins 8 and 9. The playback output signal appears at pins 6 and 11.

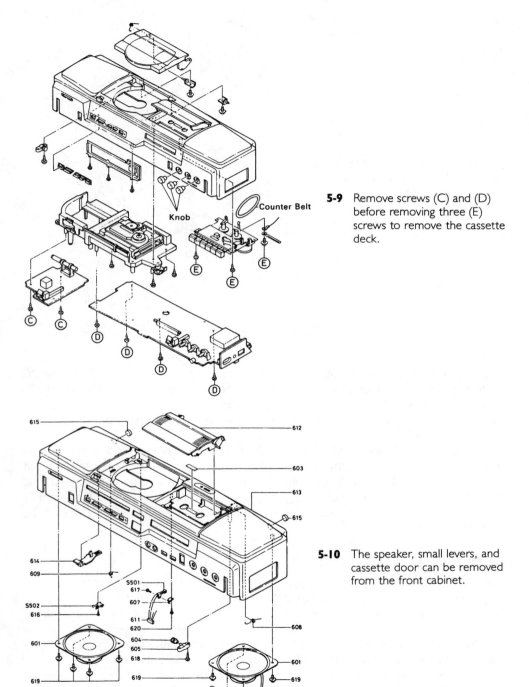

5-9 Remove screws (C) and (D) before removing three (E) screws to remove the cassette deck.

5-10 The speaker, small levers, and cassette door can be removed from the front cabinet.

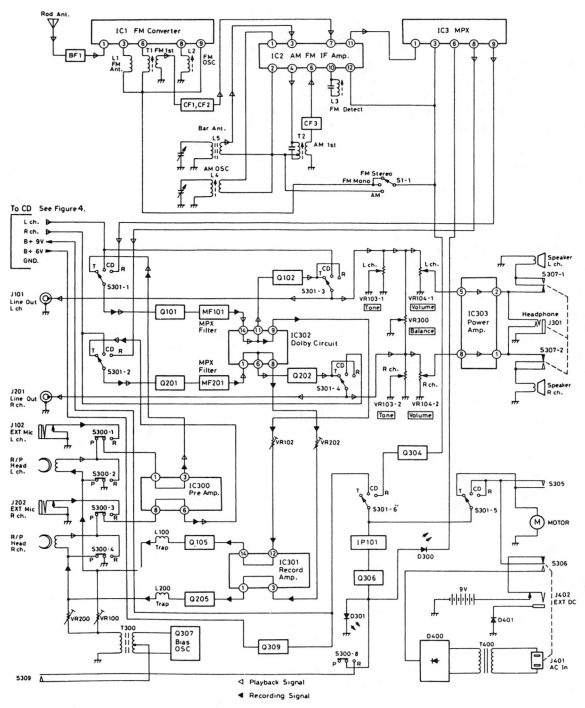

5-11 Try to isolate various troubleshooting problems with the block diagram. Radio Shack

IC302 could be defective only in one stereo record channel or in both and still not affect the playback (FIG. 5-12).

Both record and play use the preamp circuits (IC300). If the cassette player will not play or record, suspect the R/P tape heads (IC300 and IC302). Often, only one stereo channel will be weak, distorted, or dead. If both stereo channels are dead, weak, or distorted, check R/P heads (IC302 and IC303). By trying to isolate each record or playback mode with the block diagram, you can quickly isolate the defective component (FIG. 5-13).

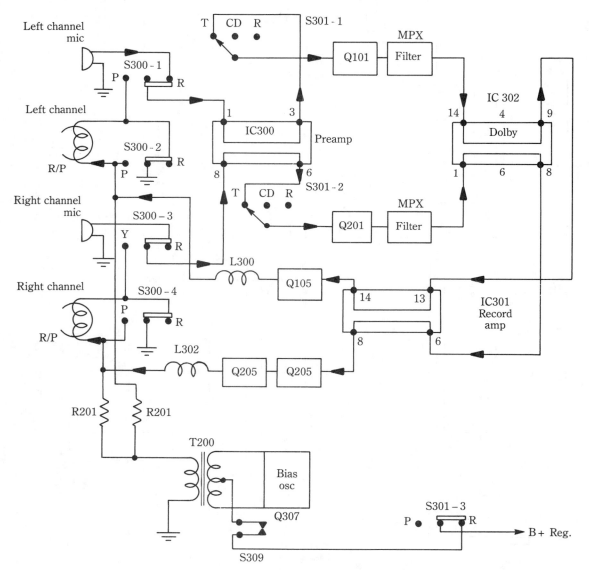

5-12 Follow the arrows in the record mode of the left and right stereo channels.

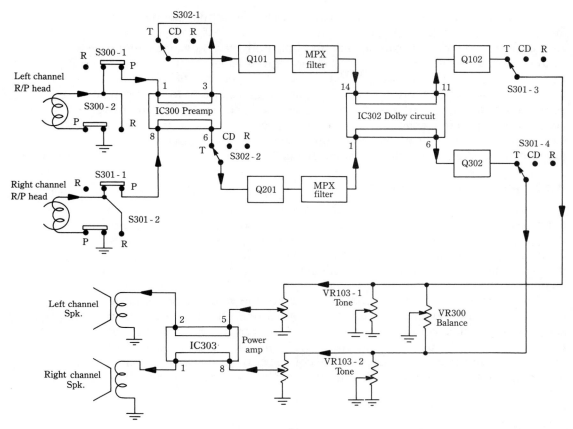

5-13 Follow the dark arrows in the stereo channels of the playback mode.

STEREO AMP CIRCUITS

One advantage in servicing stereo circuits is that the other channel can be used as a gain signal reference in each stage. If the left channel is weak and the right channel normal, use the right channel to measure the gain of each transistor or IC. Check the signal at both volume controls and compare them. If the left channel is a lot weaker, check the preamp circuits. When both channels are fairly normal at both volume controls, suspect a weak AF transistor or IC output stage.

Transistors do not become weak; however the circuit and components that are tied to the transistor or IC can create a weak stereo channel. A change in the base and emitter bias resistors can create weak audio. Leaky or open components that are tied to the IC amp can produce weak sound. Doublecheck each electrolytic coupling capacitor that is between stages. The signal should be the same on both sides of the capacitor. A low power-supply voltage can cause weak sound.

Just about any component within the stereo circuit can cause a dead channel—especially test transistors in and out of the circuit. Poor switch contacts can kill audio. Suspect an open R/P tape head when the stereo channel will not record or play. Often, a dead channel with excessive distortion or hum can occur in the audio output circuits. Feel the power output IC and notice if it is quite hot. The shorted or leaky power IC can become red hot. Both stereo channels might be dead if a common preamp or output IC component is bad.

Before replacing a leaky or open transistor or IC, check the bias resistors and bypass capacitors. The shorted IC or transistor can cause bias or load resistors to smoke and run hot. If you do not have a schematic of the stereo cassette player, compare the defective channel resistors and capacitors with the good channel. Check the operating voltages on each component after you replace all defective parts and compare them with the normal channel.

IMPROPER AUDIO BALANCE

Check the setting of the balance control when one channel is weak or down, compared to the other. If it's only slightly lower, do not worry about it. Make up the difference by adjusting the balance control. Usually, improper balance results from one stage having weak audio. Replace the erratic or worn balance control when it makes excessive noise in adjustment or when the audio cuts out in one channel.

The balance control is usually found ahead of the volume control in stereo cassettes (FIG. 5-14). When one channel will not balance, check for a loss of audio ahead of the balance control. Sometimes these balance controls can short internally to ground and cause one weak channel. Simply remove the grounded terminal of the balance control and notice if the control changes the audio. An internal broken ground terminal can cause improper balance. Replace defective balance control.

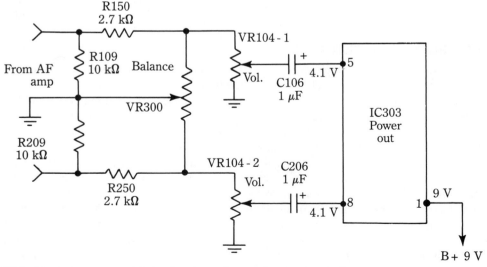

5-14 The balance control, balancing the audio to ground with VR300, is often ahead of the volume control.

GARBLED RECORDING

Notice if the garbled music is coming out of one channel or both. Is the new recording garbled? Play a commercial recording to see if the recording or the playback is causing the problem. Clean the R/P and erase heads (FIG. 5-15). Spray fluid in the record/playback switch contacts. Recheck the machine with a good cassette.

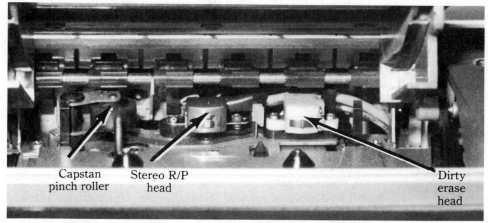

Capstan
pinch roller

Stereo R/P
head

Dirty
erase
head

5-15   Clean a dirty erase head to eliminate garbled or jumbled music.

If the garbled music is in the recorded cassette, try another new cassette and record 10 minutes of music. If the new recording remains garbled, check the oscillator bias circuits. Make sure that the bias signal is applied to each stereo R/P tape head. Check the erase head for the dc applied voltage in the record mode. Often, garbled music out of one channel indicates a dirty tape head or cassette. When garbled in both channels, check the erase and R/P heads, the bias oscillator circuit, and the defective amplifier channel.

CASSETTE DOOR WILL NOT OPEN

Suspect that tape has wrapped tightly around the capstan when the cassette door will not open (FIG. 5-16). Do not pry open the door. Remove the back cover and try to locate the capstan/flywheel. Rotate the capstan backwards by hand. This action will loosen the tightly wrapped tape so that the door can be opened.

Of course, the cassette tape might be damaged to the point where it cannot be used or broken into. Sometimes the tape can be saved if it unwinds without any problems. You will have to cut the tape to remove all of it from capstan and pinch roller. Doublecheck the bearing of pinch roller for excess tape wrapped between rubber roller and the outside bracket support. Clean the capstan, pinch roller, and spindles after the tape spills out.

If the door catch will not let the door open because it is jammed or bent, try to release the door from the back side. The cassette tape might be normal, but

5-16   The jammed cassette with tape pulled out will not let the cassette door open. Do not pry on the plastic door.

have a broken door hinge or release piece. Sometimes, the front bezel or plastic front can be removed with a couple of screws; then you can see what is causing the door to not open. A jammed gear or lever of the slow eject mechanism can prevent the door from opening.

Be careful when prying on the plastic door; don't break or throw the door out of alignment. Pry lightly on the front door and rotate the capstan backwards to release the excess tape around capstan and leave the door slightly open. Usually, the cassette tape is damaged before the door is released and the cassette is removed. It's much easier to purchase another cassette than it is to locate another cassette door.

CLEANING THE OPTICAL LENS

Be very careful when working around the CD player to avoid exposure to laser beam radiation. You can damage your eyes if you stare at the bare optical lens assembly while the player is operating, so keep your eyes away from that area. Obey the warning labels that are fastened to the laser optical assembly. Always keep a disc loaded while repairing the CD player (FIG. 5-17). Remember, the laser beam is not visible like that of an LED or pilot light. Do not defeat any interlock switches to check the laser operation.

A dirty optical lens can cause improper searching and playing of the laser assembly. Before you attempt to repair the optical sensor assembly, clean the lens area. Excess dust, cigarette or cigar smoke, stains, and tarnishes can occur on the optical lens. The laser beam can't reach the compact disc. Be careful not to apply too much pressure when cleaning the optical lens or you might damage the lens assembly (FIG. 5-18).

Clean the lens with lint-free cotton or with cleaning paper for camera lenses that is moistened with a mixture of alcohol and water. Cleaning solution for cameras is ideal. Wipe the lens gently to avoid bending the supporting spring. Blow

5-17 Always have a CD on the turntable while servicing CD player circuits.

5-18 A close-up view of the laser lens assembly. Do not stare directly at lens area.

excessive dust from the optical lens with a can of dust spray, which is available in camera departments.

CD MOTORS

In the standard CD player, a loading motor slides out to load the disc. In the portable CD/cassette player, you push a button and the top lid raises up or comes out from the front so that the disc can be loaded. The lid is pushed down or in to load the CD manually (FIG. 5-19). The top lid (655) is hinged at the back so that the disc can be placed on the turntable of the disc motor assembly.

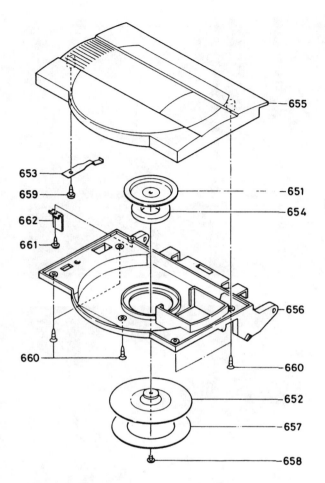

5-19 How the CD turntable and top lid are mounted in a CD case.

Disc motor

The CD motor rotates the disc at approximately 500 rpm at the inside and slows down to about 200 rpm as the laser assembly moves toward the outer rim. The disc motor is also called a *spindle* or a *turntable motor* by other manufacturers. The disc-motor shaft is directly fastened to the disc table (FIG. 5-20).

The disc-motor voltage is supplied through a driver transistor or IC. The driving transistors are operated from a servo-controller IC (FIG. 5-21). The driver voltage is zero at the emitter terminals until a signal voltage is applied to the base terminals. This voltage can be under 10 V in the disc-motor circuits.

Suspect a defective motor, drivers, and no servo signal when the disc motor will not rotate. Quickly measure the voltage across the spindle motor. If it has no motor voltage, check the voltage that is supplied to the driver transistors. Remove one lead of the disc motor and check the winding with the RX1 ohmmeter range. Replace an open motor with the exact part number.

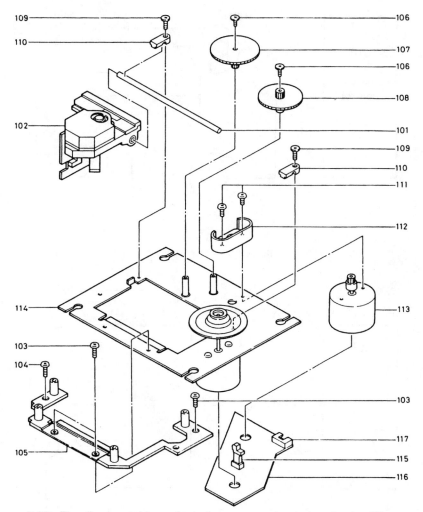

5-20 The disc, turntable, and spindle motor are located under the CD.

Sled motor

The slide, sled, carriage, or feed motors move the optical pickup assembly across the disc from the inside to the outside rim of the CD. The motor is gear-driven to a rotating gear (108 and 107 in FIG. 5-20). The gears glide the optical assembly down a sliding rod. Usually, the SLED motor is driven with a transistor or IC circuit (FIG. 5-22).

Check the slide motor-drive circuits with accurate voltage and resistance measurements. Make sure that both positive and negative voltages are present at the respective driver transistors or ICs. Take in-circuit transistor tests for each transistor with the diode transistor test of the DMM. Check the continuity of the

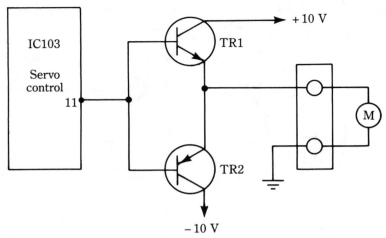

5-21 The disc motor voltage is controlled by driver transistors TRI and TR2 with servo control IC103.

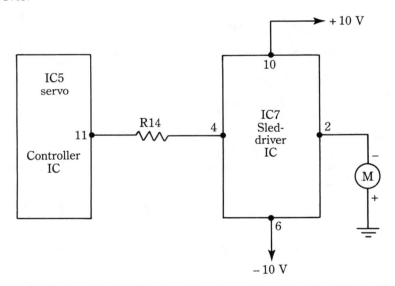

5-22 A block diagram of the SLED motor, driven by a voltage from the IC7 driver and a signal from servo controller IC5.

motor winding with RX1 ohmmeter scale. Remember, in intermittent operations, one transistor might be opening only under load conditions. It might be necessary to replace both transistors.

Improper output voltage on the motor can be caused by a defective driver IC. Make sure the positive and negative voltages are applied to the driver IC (pins 6 and 10). Monitor the voltage at the SLED motor terminals. A leaky driver IC can cause improper voltages applied to the SLED motor terminals. Replace all components with exact part numbers.

CD BLOCK DIAGRAM

Although, the servo, Dac, DSP, and laser pickup circuits are fairly complicated, many circuits can be checked in the CD player (FIG. 5-23). The disc and SLED motor circuits can easily be checked with voltage measurements on the DMM. The laser, door, and P-set switch contacts can be checked with the ohmmeter. The dc/dc converter input and output voltages, and power-supply voltages cause many problems within the CD player (FIG. 5-24).

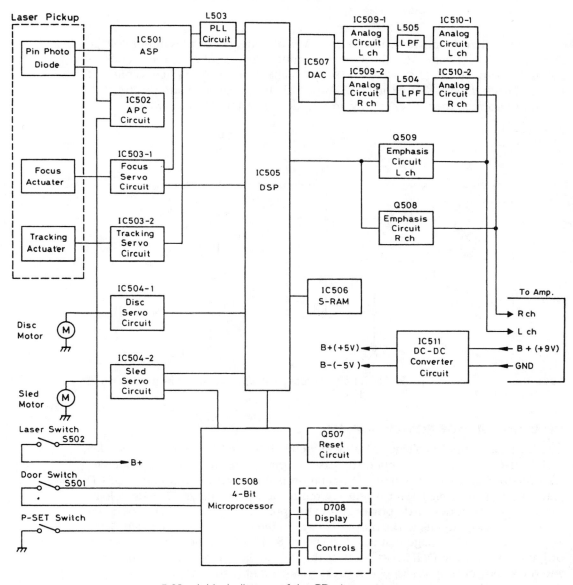

5-23 A block diagram of the CD player. Radio Shack

5-24 The +9 V from the batteries or ac power supply is fed to a dc/dc converter circuit IC511. A 5-V positive and negative voltage is applied to the CD circuits.

When either the positive or negative 5 V is missing, check the output voltage from the dc/dc converter IC. Check the B+ input voltage applied to the V_{CC} terminal of the converter IC (FIG. 5-25). The supply voltage (+9 V) is applied to pin 9 of IC511. Check the 5-V source at TP22 and TP23. If the negative and positive 5 V are found normal, the power supply and dc/dc converter circuits are okay.

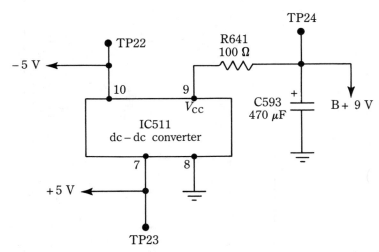

5-25 The +9 V is fed into pin 9 of IC511 and the positive 5 V at pin 10 and negative at pin 7. Check the 5-V source at TP22 and TP23.

CD LOW-VOLTAGE POWER SUPPLY

The CD power-supply voltage (+9 V) is switched into the CD circuits with S301-5 and S301-6. Check S301-6 when the cassette player and radio circuits operate. Clean S301-5 and S301-6 when CD operation is erratic or intermittent. The CD player utilizes the same power supply source as the cassette player. The dc voltage from the 9-V batteries or dc power supply is applied to S301-5. S301-6 switches the CD regulator circuits to the output voltage to the CD-player circuits (FIG. 5-26).

Check the B+ voltage that is fed to switch S301-5 from the ac power supply or batteries with the DMM (FIG. 5-27). If the voltage is low on both the batteries and the ac power supply, suspect an overloaded circuit in the CD player. Switch to the cassette player and notice if the +9 V returns. Check the CD player circuits

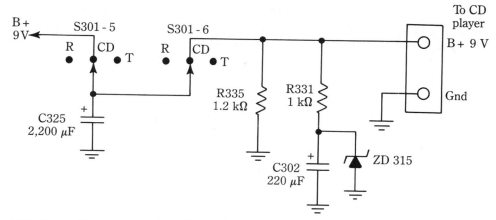

5-26 The + 9-V source must be fed through switch S301-5 and S301-6 before being regulated by ZD315.

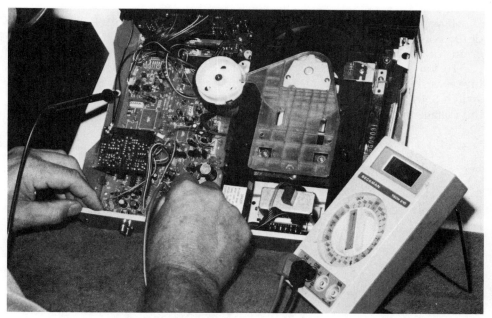

5-27 A technician checking out the B+ voltage from the power-supply circuits with a DMM.

when the voltage is low and normal on the cassette player or radio. Do not overlook the possibility that zener diode ZD315 might be leaky or R331 might be burned.

NO SOUND FROM THE CD PLAYER

The *digital-analog converter (DAC)* changes digital signals to audio signals at IC507. The analog audio signal output of IC507 is applied to analog IC509-1 and

IC509-2. The low-pass filters (L504 and L505) are found between the output IC510-1 and IC510-2. The audio signal can be traced with an external audio amp or signal traced at IC509-1 and IC509-2 to the RCH and LCH outputs (FIG. 5-28). The two-channel audio signal is fed to the audio output amp circuits.

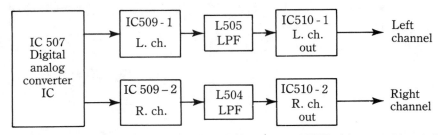

5-28 You can signal trace the audio signal from D/A converter IC507 with an outside amplifier or signal tracer.

If you hear no CD audio from either channel, suspect no B+ voltage to the dc/dc converter (FIG. 5-29). Clean switches S301-3 and S301-4 if the CD audio is erratic or nonexistent. If correct voltages are applied to IC509-1 and IC509-2, and IC 510-1 and IC510-2, suspect a defective IC. If only one channel of audio is missing or weak, signal trace the audio at the input of IC509-1 and IC509-2 through to switch S301-3 and S301-4. When the audio signal stops or appears weak, check the voltage in the defective IC circuit.

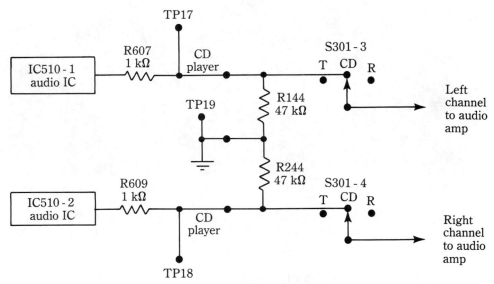

5-29 The CD stereo audio signal at TP17 and TP18 are fed to function switch S301-3 and S301-4 to the main stereo amplifier.

Both of the stereo analog IC circuits are identical and can be compared to the normal channel (FIG. 5-30). Remember, the audio signal at pin 3 of IC509-1 and pin 5 of IC509-2 is rather weak; the audio signal will increase after each IC amplifier. If the audio signal is the same at pins 3 and 5 of IC509 and weak at pin 1 of IC510-1, suspect a defective IC509-1 or a defective component in the circuit between pin 1 of IC509-1 and pin 3 of IC510-1. Also, C562 or L504 and IC509-1 could be open.

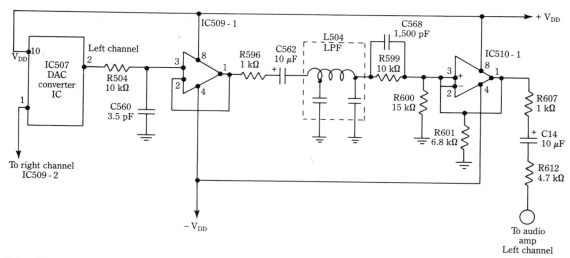

5-30   The audio signal from D/A converter (IC507) is amplified by IC509-1 and IC510-1 of the left channel. Both right and left stereo channels are identical.

Likewise, if the audio signal is the same at the output pins 1 of IC510-1 and pin 7 of IC510-2 and weak at CD player switch S301-4 of the right channel (FIG. 5-31), suspect IC510-2, R609 (1 kΩ), C575 (10 μF) and R614 (4.7 kΩ). Check the audio signal at TP17 and TP18 and see if the amount of volume is the same. If the signal is quite close, check S301-4 for dirty contacts. Always clean the function switch when either the radio, tape, or CD audio is missing.

5-31   The analog circuits of IC509 and IC510 with pin connections.

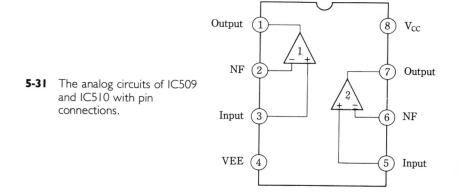

CONCLUSION

Be careful when working around the laser head assembly. Do not look directly at the laser beam assembly while operating. You can watch the laser assembly search, going up and down, from a side view. Keep the CD disc on the turntable at all times.

Check TABLE 5-1 for CD troubleshooting procedures. Figure 5-32 shows the exploded view of the cassette player cabinet and FIG. 5-33 shows the tape mechanism. Check TABLE 5-2 for troubleshooting the cassette player. Figure 5-34 shows the CD player schematic diagram and FIG. 5-35 shows the cassette player schematic with the ac power supply.

Table 5-1. The CD Troubleshooting Chart

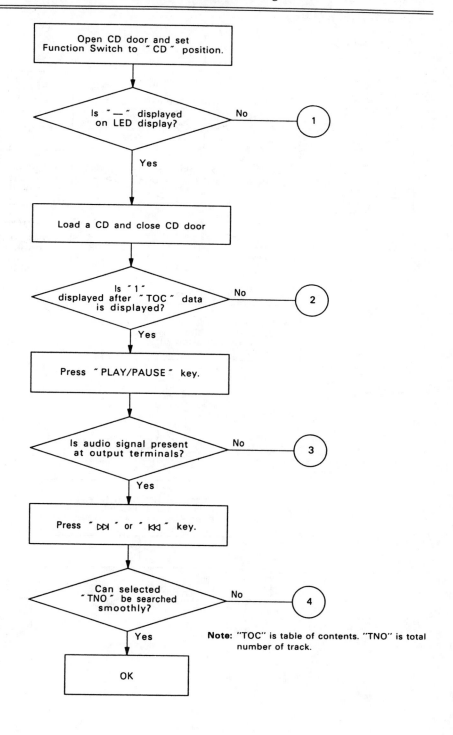

Note: "TOC" is table of contents. "TNO" is total number of track.

Table 5-1   Continued.

[Repair Item 1]

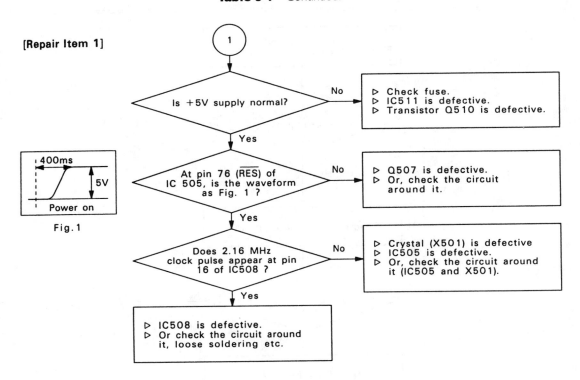

Fig. 1

[Repair Item 2]

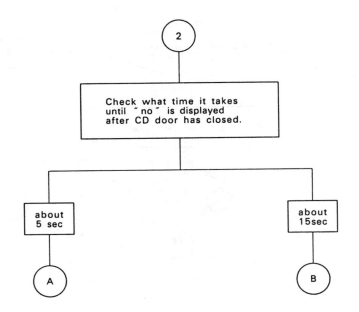

Table 5-1 Continued.

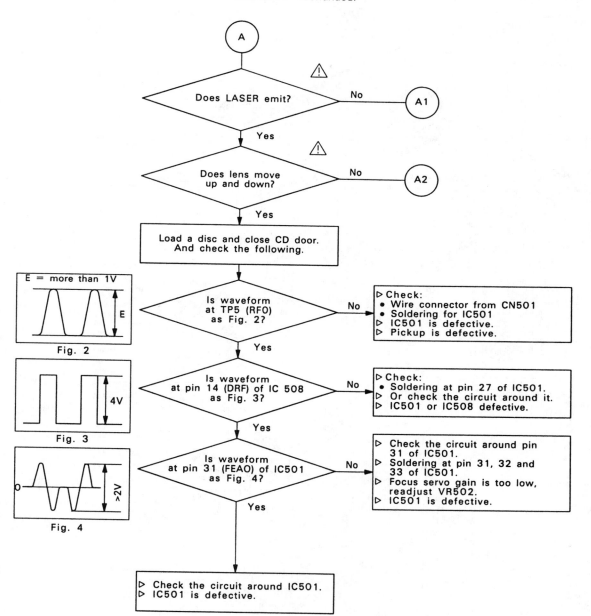

A

Does LASER emit? — No → A1

Yes

Does lens move up and down? — No → A2

Yes

Load a disc and close CD door.
And check the following.

E = more than 1V

E

Fig. 2

Is waveform at TP5 (RFO) as Fig. 2? — No →
▷ Check:
• Wire connector from CN501
• Soldering for IC501
▷ IC501 is defective.
▷ Pickup is defective.

Yes

4V

Fig. 3

Is waveform at pin 14 (DRF) of IC 508 as Fig. 3? — No →
▷ Check:
• Soldering at pin 27 of IC501.
▷ Or check the circuit around it.
▷ IC501 or IC508 defective.

Yes

0
>2V

Fig. 4

Is waveform at pin 31 (FEAO) of IC501 as Fig. 4? — No →
▷ Check the circuit around pin 31 of IC501.
▷ Soldering at pin 31, 32 and 33 of IC501.
▷ Focus servo gain is too low, readjust VR502.
▷ IC501 is defective.

Yes

▷ Check the circuit around IC501.
▷ IC501 is defective.

5-32 The exploded view of the cabinet with CD and cassette components. Radio Shack

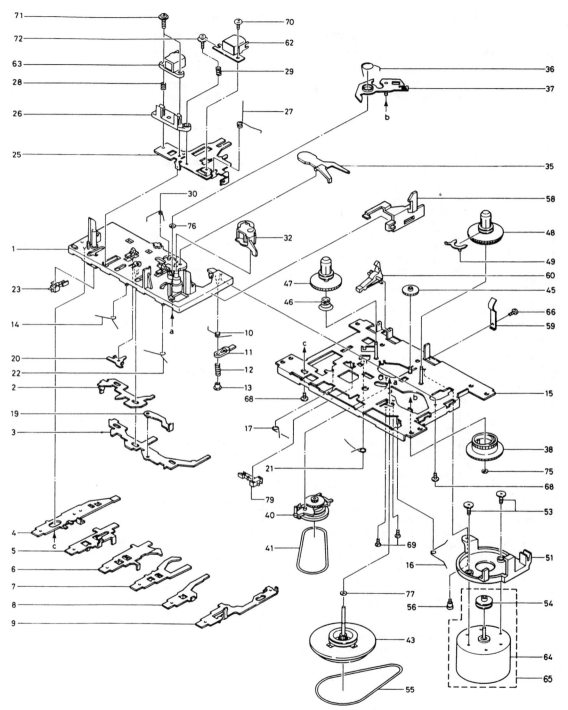

5-33 The exploded view of the tape mechanism of the cassette player. Radio Shack

Table 5-2. The Radio and Cassette Troubleshooting Chart

Symptom	Cause	Remedy
Output level too low **Overall:**	Power amplifier circuit 1. Faulty IC303. 2. Shorted C131 and C231.	1. Check and replace 2. Check and replace.
Tape:	Pre-amplifier circuit Faulty IC300.	Check and replace.
FM:	FM Front/End circuit 1. Faulty IC1. 2. Open or shorted C2. 3. Open or shorted VC.	1. Check and replace. 2. Check and replace. 3. Check and replace.
	FM decode circuit 1. Faulty IC3. 2. Shorted C32, C33, C34, C35, R26 and R27. 3. Faulty C36 and/or C37.	1. Check and replace. 2. Check and replace. 3. Replace the faulty capacitor.
AM/FM:	FM-AM IF amplifier circuit Faulty IC2.	Check and replace.
Poor tape high frequency response	1. Incorrect head azimuth. 2. Faulty REC/PB head (62). 3. Shorted C100 and C200.	1. Adjust head azimuth. 2. Clean and replace REC/PB head. 3. Check and replace.
No sound	Power supply circuit 1. Faulty S301 and S306 or poor contact. 2. Faulty J401 and J402 or poor contact.	1. Check and replace. 2. Check and replace.
	Power amplifier circuit 1. Open or shorted IC303. 2. Open or shorted C131 and C231.	1. Check and replace. 2. Check and replace.
	Output circuit 1. Open or shorted speaker voice coil. 2. Faulty J301 or poor contact.	1. Check and replace. 2. Check and replace.
No sound	Pre-amplifier circuit 1. Open or shorted IC300. 2. Open or shorted REC/PB head (62). 3. Open REC/PB head leads. 4. Open VR104.	1. Check and replace. 2. Check and replace. 3. Check REC/PB head leads. 4. Check and replace.
	FM tuner circuit 1. Faulty S1 or poor contact. 2. Faulty IC1.	1. Check and replace. 2. Check and replace.
	AM converter circuit 1. Open or shorted L4, L5 and/or T2. 2. Shorted PVC.	1. Check and replace. 2. Check and replace.

Symptom	Cause	Remedy
	AM/FM IF amplifier 1. Faulty IC2. 2. Open or shorted L3. 3. Shorted CF1, CF2, and/or CF3.	1. Check and replace. 2. Check and replace. 3. Check and replace.
Tape inoperative	1. Motor (64) dead. 2. Capstan belt (55) slipping. 3. Leaf switch (S305) poor contact.	1. Check motor lead-wires and re-placed motor. 2. Wipe flywheel (43) and replace capstan belt. 3. Adjust or replace leaf switch.
Won't take-up tape	Capstan belt (55) slipping.	Wipe flywheel (43) and/or replace capstan belt (44).
No fast-forward and rewind	Clutch assembly (40) slipping.	Wipe flywheel (43), clutch assembly (40), and/or replace FF/rewind belt (41).
Excessive wow	1. Motor (64) defective. 2. Pinch roller (32) dirty.	1. Replace. 2. Clean or replace.
Uneven speed	1. Motor (64) defective. 2. Motor pulley (54) slipping. 3. Capstan belt (55) slipping.	1. Replace. 2. Adjust or replace motor pulley. 3. Wipe flywheel, motor pulley and replace capstan belt.
No playback	1. REC/PB head (62) defective or open. 2. REC/PB head dirty. 3. Open or shorted REC/PB head leadwires. 4. No power to amplifier. 5. Defective component(s) in amplifier.	1. Replace. 2. Wipe REC/PB head with a cloth moistened with alcohol. 3. Replace wire. 4. Replace leaf switch (S305). 5. Check and replace defective component(s).
Low playback or distorted playback output	1. Amplifier defective. 2. REC/PB head dirty. 3. REC/PB head badly worn.	1. Check and replace defective component(s) 2. Wire REC/PB head with a cloth moistened with alcohol. 3. Replace.
No erase	1. Erase head (63) defective or open. 2. Open or shorted erase head lead-wires.	1. Replace. 2. Check and replace.
No record	1. REC/PB head (62) defective or open. 2. Component(s) in amplifier defective. 3. REC/PB head dirty.	1. Replace. 2. Check and replace defective component(s). 3. Wipe REC/PB head with a cloth moistened with alcohol.

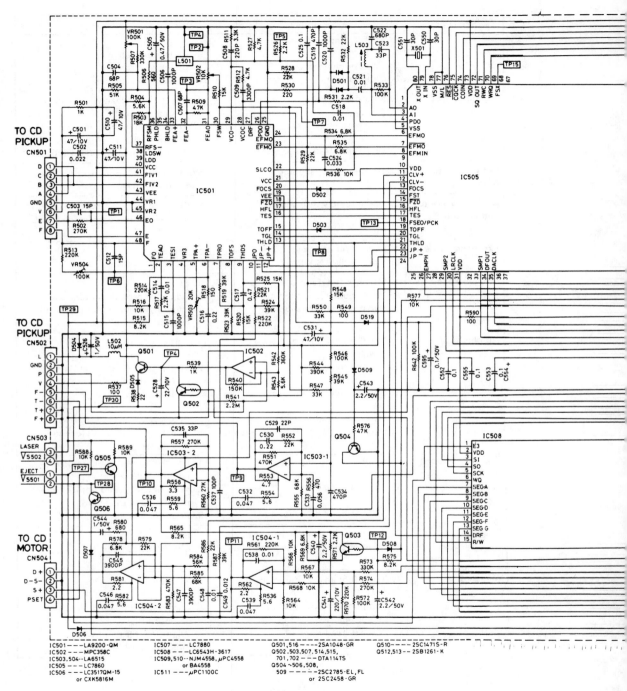

5-34 The complete CD schematic diagram. Radio Shack

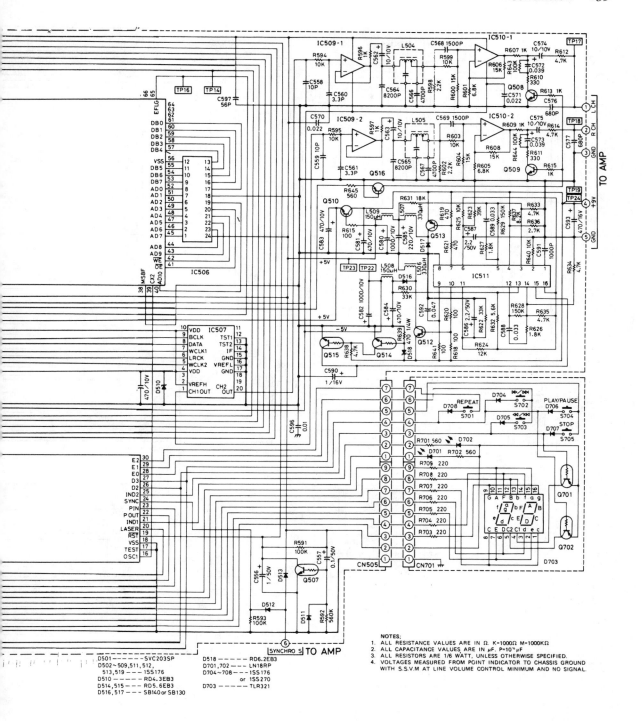

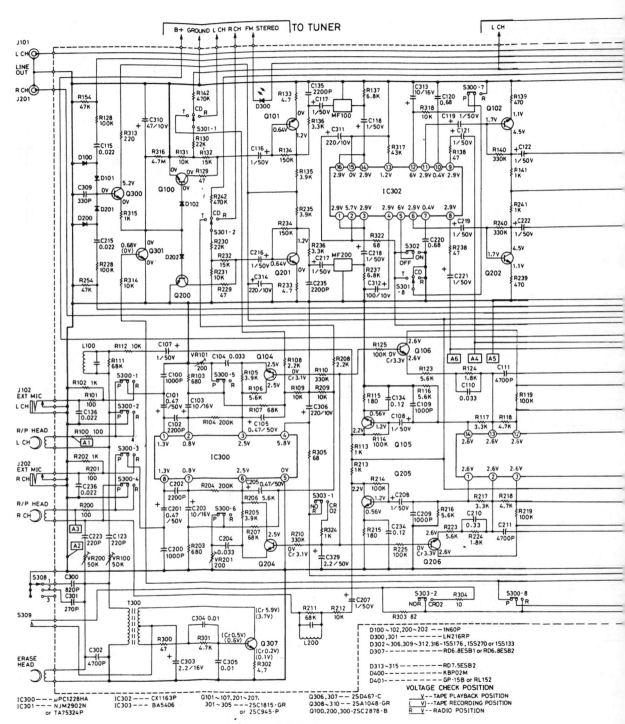

Table 5-34 Continued.

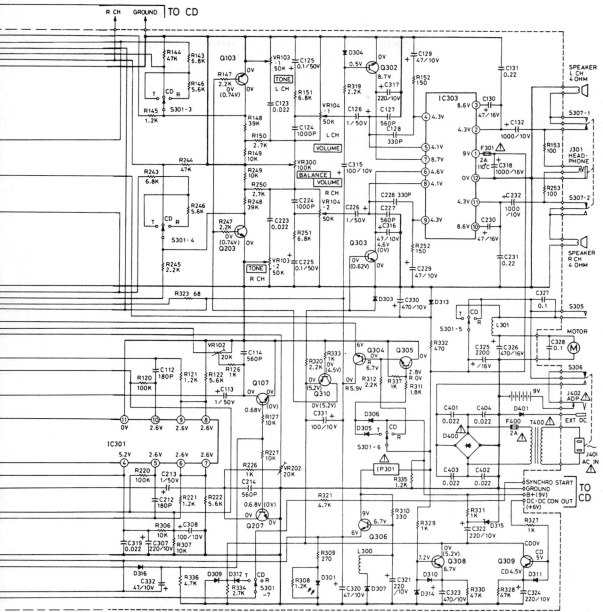

S300 -- RECORD/PLAYBACK SWITCH (PLAYBACK POSITION)
S301 -- TAPE/CD/RADIO SWITCH (CD POSITION)
S302 -- DOLBY NR ON/OFF SWITCH (OFF POSITION)
S303 -- NORMAL/CRO2 SWITCH (NORMAL POSITION)
S305 -- TAPE ON/OFF SWITCH (OFF POSITION)
S306 -- AC/DC SWITCH (DC POSITION)
S307 -- SPEAKER/HEADPHONE SWITCH (SPEAKER POSITION)
S308 -- BEAT CUT 1/2/3 SWITCH (1 POSITION)
S309 -- REC BIAS CUT SWITCH (OFF POSITION)

CD V -- CD POSITION
Cr V -- CRO2 POSITION

NOTES;
1. ALL RESISTANCE VALUES ARE IN Ω. K=1000Ω M=1000KΩ
2. ALL CAPACITANCE VALUES ARE IN μF. P=10⁻⁶ μF
3. ALL RESISTORS ARE 1/6 WATT, UNLESS OTHERWISE SPECIFIED.
4. VOLTAGES MEASURED FROM POINT INDICATOR TO CHASSIS GROUND
   WITH S.S.V.M AT LINE VOLUME CONTROL MINIMUM AND NO SIGNAL.

Chapter **6**

Repairing microcassette recorders

*M*ost microcassette recorders operate from batteries for portability, but the ac adapter can be used on some models. These units require two AA batteries. In some units, nicad batteries, and ac chargers/adapters are used. These small recorders use a total of three bolts (FIG. 6-1).

The compact and easy-to-operate microcassette recorder can have a VOR system. This voice-operated recording system is economical for tapes, batteries, and operation time. You can start recording directly from the playback mode in several units. This function is convenient to correct a previously recorded portion.

Some models have an automatic shutoff mechanism. In recording or playing, the tape stops at the end and the locked button will be released automatically (automatic shutoff). Besides a cue (fast forward) and review (rewind) control, some have a pause switch. The stop/eject button completely stops the motion and ejects or raises the microcassette. To stop the tape momentarily in the playback or recording mode, slide the pause switch in direction of the arrow (FIG. 6-2).

The deluxe microcassette recorder can have two speeds 2.4 cm and 1.2 cm. The 2.4-cm speed is recommended for normal use. A 60-minute recording can be made using both sides of the MC-60 microcassette. A built-in tape counter and instant edit features are used in some models. A full-featured recorder might have auto reverse and a speaker.

Although most units have a built-in condenser microphone, the ultra-compact recorder might have a detachable external microphone system. This small microphone is usually found at the top end of the recorder.

MICROCASSETTES

The microcassette measures only $1^1/_4 \times 2$ inches and they fit in only the microcassette player/recorder. These cassettes can be purchased in a 4-pack plastic case

6-1 The small micro-cassette recorder might operate from two small AA batteries.

6-2 To pause momentarily, place the pause switch in the direction of the arrow.

(FIG. 6-3). The standard microcassette has a small indentation on the top side of the plastic case. The nonstandard microcassette will not play in the standard recorders because their ''L'' dimension is different (FIG. 6-4).

The MC-60 microcassette will play for 60 minutes by using both sides at the 2.4-cm speed. Switch the tape speed to 1.2 cm for 120-minute operation. To prevent the cassette from accidentally being erased, break out plastic tab at end of the right-hand top side. To reverse the process, cover the broken-out slot with tape. Remember, when a certain cassette will not record, check for broken tab at the end of the cassette.

6-3 Microcassettes can be purchased in packs of four.

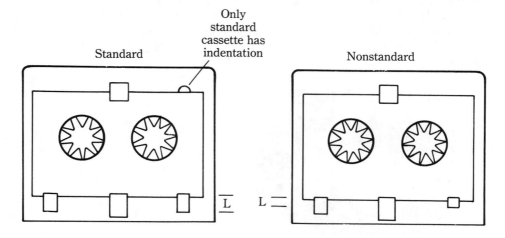

Only standard cassette has indentation

Standard

Nonstandard

App. 5 mm

App. 2.5 mm

6-4 The correct dimensions of the cassette indicates if it is a standard or nonstandard cassette.

TO RECORD WITH BUILT-IN MICROPHONE

Set the desired tape speed (2.4 cm) for normal use. Insert the cassette correctly. If the microcassette recorder has VOR recording, switch VOR to the L position (recommended for dictation). Switch the VOR off, if you don't want to use that operation. Push in the record switch and start recording (FIG. 6-5). Notice that when it is in VOR operation, the cassette will rotate only during periods of noise or talking.

6-5 The record switch is located on the right side of this Sony M-440 V VOR Pressman microcassette (TM) recorder.

For quick review, simply push the cue/review control toward review while recording. When the control is released, it begins to play back the recorded material. On some models, you can listen to the sound being recorded through the earphone jack. If you want to stop the tape momentarily when either playing or recording, slide the pause switch in the direction of the arrow.

PLAYBACK MODE

The same tape speed must be used for playing and recording. If not, the recording will sound too fast or too slow. To hear what you have recorded, place cue/review switch to the rewind position (review). To rapidly advance the tape, slide the cue/review control to the cue position. To play the tape, simply depress the play button (FIG. 6-6).

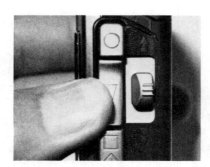

6-6   To play, press the play button.

To skip over or repeat a certain portion during playback, keep the cue/review control pushed up or down with the button depressed. When the control is released, playback begins. After advancing or rewinding the tape until its end, be sure to press the stop button to release the cue/review switch.

VOR RECORDING

VOR recording starts when you begin to talk and will stop recording automatically when you stop talking. In fact, a tap of a pencil or finger can trigger the VOR when it is set in the high-level recording position.

For VOR operation, place the VOR in the high-level position (L). Some microcassette recorders have H, L, and off positions (FIG. 6-7). Push the record button; the recorder will start and then stop when no noise or talking is detected. Start to talk and notice how the voice recording operates.

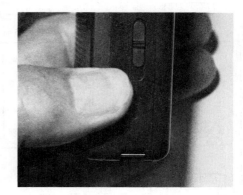

6-7 Most standard VOR recording operation is done in the high-level position.

To play a recording, make sure the volume is up. Most recorders have a recording level adjustment, built into the unit. The volume control can be in any position for recording, but it must be turned up in playback mode.

CIRCUIT BLOCK DIAGRAM

The recording and playback circuits might use two or three ICs and several transistors. Only transistors were used in early microcassette recorders. Today, you can find IC processors and transistors with surface-mounted components. Surface-mounted transistors, IC processors, capacitors, and resistors are ideal to fit into a small hand-held recorder.

The block diagram of the small microcassette recorder is rather simple with only a few components (FIG. 6-8). The solid black-and-white arrows indicate the signal paths when recording or playing. IC (U101) contains the play/record switch, automatic gain control, reference voltage, VD, audio preamp, audio power output, and alarm circuits. U101 can contain a 24- or 30-terminal flat surface-mounted component with wing-tip terminal connections.

U104 is the system-control IC that controls record/play operations, motor operations, the VOX, the pause LED, and power and voltage regulators. U103 is an 8-pin motor-control surface-mounted IC. The recording bias, reel operation, battery LED circuits, power control, and the VOX are controlled with separate transistors.

First, determine where the trouble might be on the block diagram, then locate the same section on the schematic. Remember, different voltages can be measured when in different modes. Usually, this occurs in play and record.

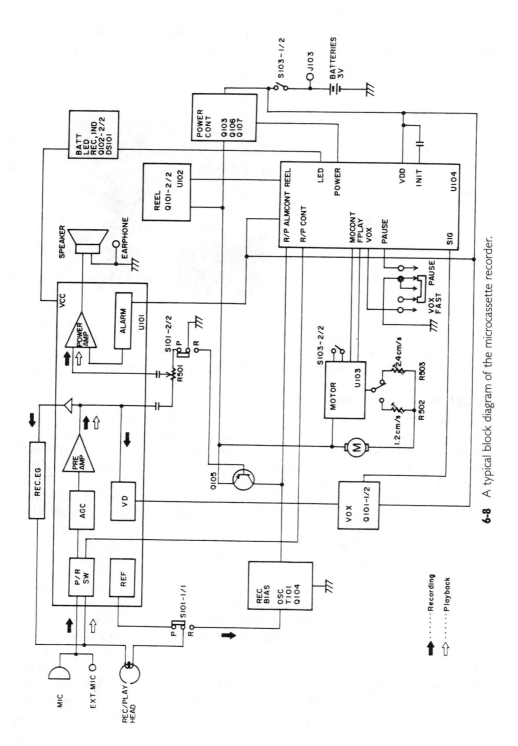

6-8 A typical block diagram of the microcassette recorder.

SURFACE-MOUNTED COMPONENTS

Surface-mounted components are tiny in size compared to regular parts found in the cassette player. Most of these parts are mounted flat against the PC board. Besides IC parts, the transistors are three-legged components with flat leads to solder on the copper foil (FIG. 6-9). Notice that two diodes in one component look like a regular transistor part.

IC AND TRANSISTOR LEAD IDENTIFICATION

6-9 Surface-mounted IC and transistor lead identification in some microcassette recorders.

Besides solid-state components, resistors and capacitors are contained in small chunks or rectangular objects. Often, small capacitors are larger than resistors. The contact of each surface-mounted capacitor or resistor is located at each end. The tinned end piece is soldered directly to the PC board wiring. Usually, these small components are damaged when they are removed from the circuit (FIG. 6-10).

Check and doublecheck each suspected component before deciding if it is defective. Make sure that the component is defective before removing it from the circuit. Some larger ICs are glued into position when they are manufactured. Often, both sides of the PC board have foil wiring to connect all components in the record/playback circuits.

DISASSEMBLY

To get at certain components for voltage, resistance, and inspection checks, the bottom cover must be removed. Remove the three bottom screws to remove the bottom cover in this microcassette (FIG. 6-11). Place the flat pan screws in a small dish, saucer, or lid so that they will not get misplaced. Remove three small Phillips screws to remove the bottom cover in the Sony Pressman model M-440V.

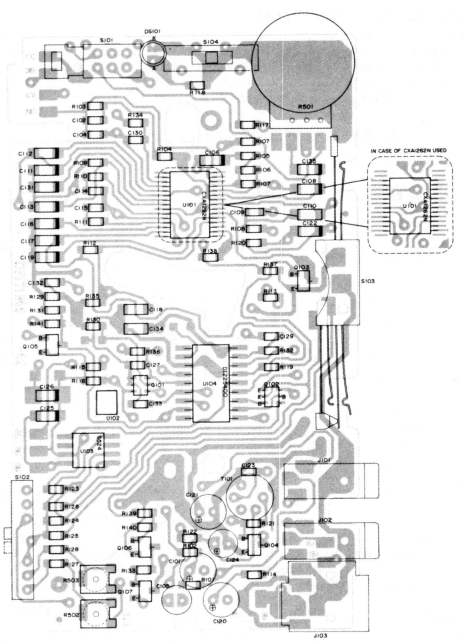

6-10 The enlarged view of the PC board with the top view showing surface-mounted components.

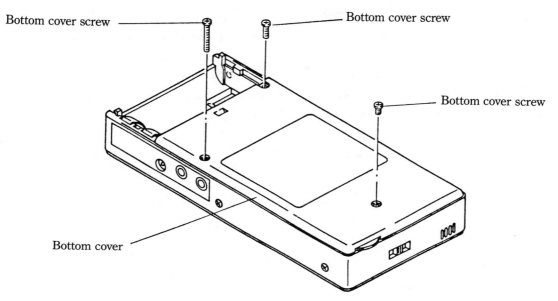

Bottom cover screw

Bottom cover screw

Bottom cover screw

Bottom cover

6-11 Remove three screws to remove the bottom cover in this microcassette recorder.

Remove the five upper cover screws in the cassette player body to remove the top cover (FIG. 6-12). This action will release the whole inside body of the recorder. In some models, after the bottom cover is removed, the main chassis can be removed by taking out several screws that hold the chassis to the top cover assembly.

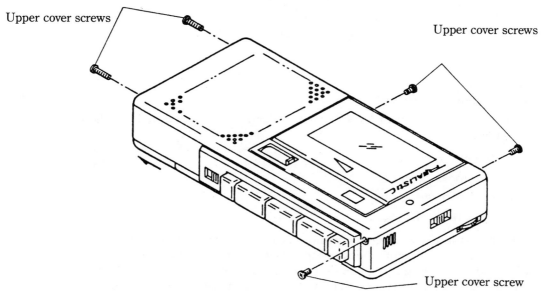

Upper cover screws

Upper cover screws

Upper cover screw

Upper cover screw

6-12 Remove the five upper-cover screws to remove the top-side cover.

Removing the PC board

Remove the bottom cover. Remove the two amplifier screws. Disconnect the following wire leads (FIG. 6-13):

1. Disconnect the two head leads (brown and black).
2. Remove three motor leads (red, black, and purple).
3. Remove two battery leads (red and black).
4. Disconnect the two speaker leads (green).
5. Remove the two mic leads (red and white).

Amplifier screws

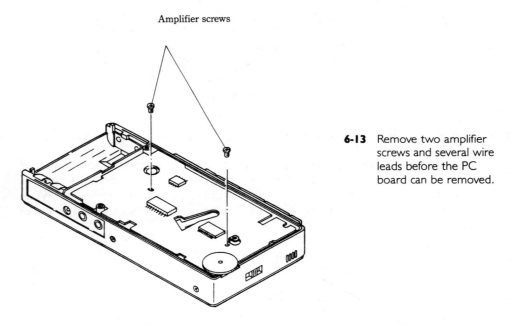

6-13 Remove two amplifier screws and several wire leads before the PC board can be removed.

Removing the cassette deck

To remove the cassette deck from the cabinet and the PC board, remove the following (FIG. 6-14):

1. The bottom cover.
2. The upper cover.
3. The PC board.
4. The cassette screw.

If you do not have a service manual or instructions on how to remove the various assemblies, carefully peek at each assembly to see what screws are holding it in position. Do not remove more parts than are needed. Study the layout and proceed. It's best to lay out all removed parts in a line as they are removed so that you can easily reverse the procedure.

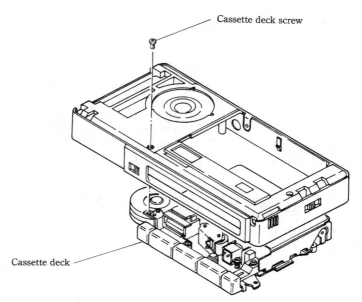

Cassette deck screw

Cassette deck

6-14　Remove one cassette deck screw so that the cassette deck can be removed.

ONLY ONE SPEED

When only one speed works in a 2-speed recorder, suspect a dirty speed switch. Sometimes, the speed is controlled by inserting a different size idler wheel. In other models, the speed is controlled with a speed regulator. Make sure the speed is not a mechanical problem. Clean all idlers and pulleys with alcohol and a cleaning stick.

If only one speed is still working after cleaning, apply cleaning fluid into the contact areas of the speed switch. With no speed or when both speeds are incorrect, check the voltages within the motor regulator circuits (FIG. 6-15).

Here, the 2.4-cm and 1.2-cm speeds are changed with switch S102. Dirty switch contacts can cause no speed or erratic speed operation. Check the critical voltages on Q102 and U103. Test Q102 with a transistor tester or with the diode-transistor test of the DMM. A leaky Q102 can produce fast speeds with S102 in any position.

When the supply voltage at Q102 and U103 are normal, suspect a change of resistance in the collector circuits or suspect a defective motor. Check the voltage across the motor terminals and notice the change when both speeds are switched into the circuit. If you find a voltage change, but no speed change, suspect a defective motor.

The tape speed of the microcassette recorder can be checked with a tape-speed cassette (W411). Adjust R503 for correct 2.4-cm/sec speed with a 3,000-Hz test tape. R502 is adjusted for the 1.2-cm/sec 1,500-Hz test cassette. R502 and R503 are located close to the tape speed switch on the Radio Shack Micro-26 model (FIG. 6-16).

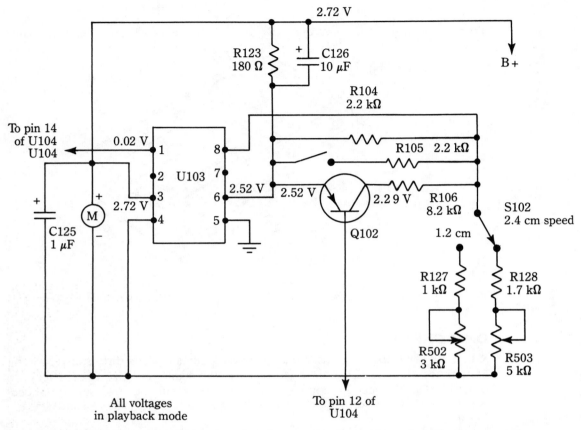

6-15 Check the speed circuits when the microcassette recorder is running at improper speed.

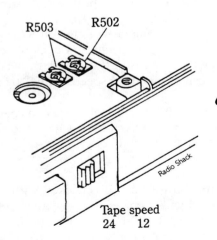

6-16 The two speed-control adjustment resistors are near the speed switch in a Radio Shack Micro-26 recorder.

WEAK BATTERIES

Loss of speed and audio can be noticed when the total voltage drops to 2.5 volts in the recorder. Check those batteries for slow speed and weak audio. If a battery tester is not handy, check the total voltage across the batteries with recorder in playback mode. Voltage measurements cannot be accurate unless the batteries are checked under load (FIG. 6-17).

6-17 Check the total voltage of both batteries with the cassette recorder operating with the DMM.

It's best to replace both batteries for best performance. Sometimes, one battery can become defective before the other. Check the total voltage across the battery terminals of each battery to locate the defective one. The batteries can be replaced by sliding open a plastic cover.

Replace the AA batteries with alkaline or nicad batteries. Of course, both batteries are heavier than regular batteries. Within the pocket microcassette recorder, extra weight can become a problem, so why not install alkaline batteries?

Doublecheck the polarity of the batteries when removing the old ones. Look into the battery compartment for positive (+) or negative (−) polarity signs. Wipe the ends of the new battery on a paper towel or a cloth towel before installing. If the recorder does not operate, quickly shut off the unit and recheck for correct battery polarity.

RECHARGEABLE BATTERIES

Nicad batteries will last a lifetime if properly charged and cared for. Although the nicad batteries initially cost more, they can last the lifetime of the microcassette recorder. Some microcassette recorders are equipped with nicad batteries and an ac charger. Of course, the recorder should be operated on the ac adapter when on dictation or when ac power is available.

If the recorder is only battery operated, you can still use nicad batteries by purchasing a nicad charger. Nicad batteries should be charged when they are weak and will not operate the cassette player. Follow the instructions found on charger. Discharge the nicad batteries when they start to operate for only a short time.

SLOW SPEED

Check the batteries for slow tape speed. If batteries are used quite rapidly, check the current drain of the recorder in operation. Just slip a piece of light cardboard between one battery and the terminal slip. Set the DMM to the 200-mA meter scale. Touch the end of battery to the battery terminal. Place the meter in series with batteries and recorder circuits. Another method is to check the current by placing meter probes across the on/off switch with the switch in the off position.

If the current is quite high, suspect a leaky amplifier or power circuit. The normal current of the Sony M-440V recorder is around 90 mA. If the current is over 125 mA, the unit might be defective.

Insert a new cassette and notice if the speed changes. Sometimes a defective cassette will drag down the play/record speed. Notice if the current fluctuates with the old or defective cassette.

Slow speeds can be caused by an overly large motor belt. Check the belt for oil spots or shiny signs of slippage. Replace the loose motor belt. Check for a dry or binding capstan wheel. If the capstan is sluggish, remove and clean it with alcohol and a rag. Place a drop of oil on the capstan flywheel bearing and replace the belt.

DAMAGED CASSETTES

Although the microcassette is small and fairly well constructed, a defective cassette can cause slow speeds, noisy operation, sound dropouts. A cassette that is wound too tightly can cause slow speeds. Noise that is directly from the cassette can be caused by dry bearings (plastic against plastic). Dropouts can occur when the tape spills out of the cassette and becomes tangled around the capstan or pressure roller. The wrinkled tape is damaged and can't record or play as it was recorded.

Replace the defective cassette with a new one. Do not try to repair the broken tape. Excess tape spilled out of the center slot can be rewound by placing pencil inside the take up reel slot and rewinding (FIG. 6-18). If it is kinked excessively, replace the cassette. Check the plastic sides for a cracked case when the hubs will not rotate.

6-18 Check for wrinkled or excess tape at the center of the cassette. Take up the slack with a pencil in one of the reel hubs.

NO REVIEW OR CUE

The review control rewinds and cue fast forwards the tape. First, determine if the tape is normal in the playback mode. If it sounds normal, the motor belt and motor are good. When the fast forward is slow, suspect that the idler wheel or reel is slipping. In many units, a larger wheel or pulley is placed against the idler and reel for faster speeds (FIG. 6-19). Clean all moving surfaces with alcohol and a stick. Check for oil on the running idlers and pulleys.

Idler wheel or gear assembly

6-19 The idler wheel can be pressed against the hub or spindle for fast forward or rewind modes.

In some cassette players, the rewind operates at the same speed as fast forward. The idler wheel is just pressed against the reel in the opposite direction, which changes the direction. Often, with this type of operation, slippage is found in both the review and cue operations. Replace the idler or pulleys if they are worn or deformed, which produces slow speed.

NOISY EARPHONE JACK

The earphone jack is usually wired in series with the small speaker voice coil. When the earphone is plugged into the jack, the speaker is cut out of the circuit (FIG. 6-20). In this circuit, the earphone sound-dropping resistor and isolation capacitor (R114 and C120) completes the circuit to common ground. In some recorders, you can listen to what is being recorded with the earphones.

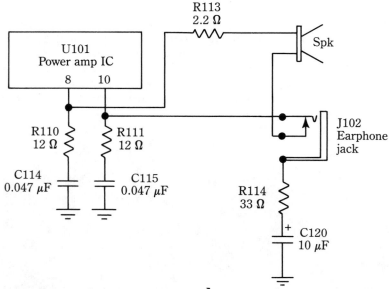

6-20   The earphone jack is in series with R114 and C120 to common ground.

The dirty earphone terminals can cause erratic sound or deaden the sound to the earphone. Within the stereo earphone jacks, one audio channel might be dead as a result of dirty or bent earphone terminals. Remove the cover and squirt cleaning fluid into the jack area. Insert the earphone jack a few times to help clean the contacts. If the earphone jack is molded, apply liquid into the jack opening. When these contacts are broken or extremely worn, the sound is intermittent or erratic. Replace the defective earphone jack.

ACCIDENTAL ERASE

Break out the small plastic tab at the right side end of the microcassette to prevent accidental erase. Break out both tabs when both sides are to be saved. If not, you can accidentally erase valuable information, before you realize it (FIG. 6-21).

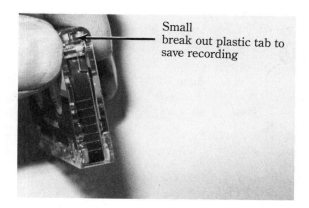

Small
break out plastic tab to
save recording

6-21   The pen points at the small plastic tab, which helps to protect the recording on the cassette.

Remember that when a microcassette will not let the record button press inward, the small plastic slot can be knocked out. Always play the cassette to make sure the recording doesn't need to be saved. Apply a small piece of tape over the opening if you want to record over this cassette.

WORKS WITH BATTERIES, BUT NOT WITH AC

A lot of microcassette recorders or players will operate from the power line with an ac adapter. The dc voltage male plug is inserted into a small jack at the end of side of the recorder (FIG. 6-22). The internal batteries are disconnected when the adapter plug is inserted.

dc
Voltage plug

6-22   The dc voltage male plug from the ac adaptor will plug into the side of this Sony microcassette recorder.

The male dc plug from the adapter breaks the circuit and switches the batteries from the circuit. Often, the dc voltage from the ac adapter is higher than normal when it is measured out of the circuit (FIG. 6-23). Inspect the switched terminal connections. Clean the contacts with cleaning fluid. Apply cleaning fluid to the soft cloth and wipe the male plug for a good contact. Replace the small dc plug, if it makes a broken or poor contact.

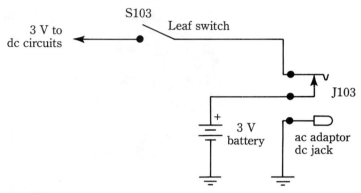

6-23 When the ac adaptor plug is inserted, the internal batteries are removed from the circuit.

To determine if the plug is defective, plug the dc male plug into the dc jack and measure the voltage at the plug terminals. If no voltage registers, the contacts might be bent or worn. Check the voltage at the male plug to determine if adapter or the dc jack is defective. If in doubt, clip the male plug connections to the dc jack terminals with small clip leads. Try to repair the ac adapter with no voltage at the male plug. Doublecheck the dc voltage polarity at the dc jack.

INTERMITTENT VOR OPERATION

An intermittent voice-operated cassette player can be difficult to repair. Notice if the tape will operate in VOR position when the VOR switch is pushed or pressed to one side. This can be rather tricky because handling of the switch or case can trigger the VOR circuits. Just keep talking while working with a suspected dirty switch.

Spray cleaning fluid in the switch area. Remove the back cover. Insert the small plastic tab into the switch area. Only a squirt or two is needed. Too much spray can drip on the drive belts or moving pulleys. Wipe off all excess with alcohol and a cleaning stick. Work the switch back and forth to help clean the contacts. Recheck operation of VOR switch after clean up.

NO PLAYBACK

Usually, when fast forward is not working, the playback is slow or the tape doesn't rotate. Check for weak batteries and replace them, if necessary. If the batteries do not last very long, suspect the overload circuits. Take a quick current check.

Check to see if recorder is in pause. Make sure the pause control is off. Do the spindles rotate with a new cassette or without? Insert a new cassette. Listen for the tape or motor to rotate. Often, by holding recorder up to your ear, you can hear the motor rotate. If the motor does not rotate and new batteries and a cassette are installed, remove the bottom cover and check the voltage that is applied to the motor terminals.

If no voltage is at the motor, the leaf switch, regulator IC, or transistor might be bad. Check the voltage at the leaf switch. No voltage at the leaf switch might indicate a defective switch or dirty contacts. Clean the contacts with cleaning fluid and a cleaning stick (FIG. 6-24).

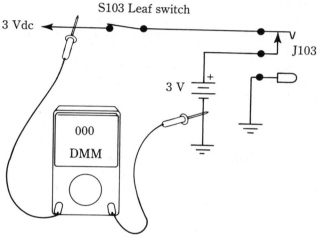

6-24 Check the leaf switch for dirty contacts or broken wire terminals if no voltage is applied to the motor circuits.

If voltage is found at the voltage regulator input and the motor doesn't rotate, suspect a defective regulator transistor. Measure the voltage at the IC regulator. The regulator can be defective if the input voltage is measured with no output voltage. Check all resistors within the motor circuits. If voltage is applied to the motor terminals and the motor doesn't rotate, replace the defective motor. These motors must be replaced with motors with the exact part numbers.

Suspect a broken motor belt or binding capstan wheel when the motor is rotating, but the tape isn't (FIG. 6-25). Check the play spindle idler wheel when the capstan is rotating. Do not overlook a locked brake lever that does not release. Often, when the brake or capstan wheel prevents tape motion, the motor pulley is rotating inside the motor belt. Sometimes, a burned rubber idler or spindle surface will have a slot burned in it, which prevents tape action.

6-25   Check for a broken or loose motor drive belt if the motor is operating and the tape is not moving.

NO RECORDING

First, check the microcassette for the removed plastic tab in the right end of cassette. The recording button cannot be pushed in when this tab is removed. Try another cassette, and clean the R/P tape head.

Does the recorder play other cassettes? If so, the head playback and recording circuits are normal in the path of audio signal. Check the bias signal with a scope at one side of the tape head in the recording mode. If a scope is not handy, check the voltages on the bias transistor. It might help to clean the record/play switch.

The bias exciting voltage from the bias oscillator is applied to the record/play tape head through S101 (FIG. 6-26). Practically no voltage (0.06 V) is found on the base and collector terminals of bias oscillator Q104. Check the bias voltage between the base and the emitter (0.32 V). If the collector voltage is low, suspect a leaky Q104 or an insufficient supply voltage.

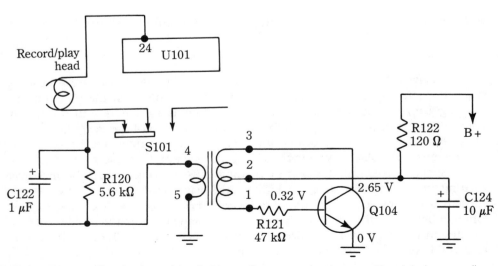

6-26   The recording circuit consists of a bias oscillator to excite the record head during recording.

Take resistance measurements of both the primary and secondary windings of T101. Check for poorly soldered contacts around terminals 1, 2, and 3, for erratic recording. Measure the resistance from one side of the tape head to the chassis ground (5.6 kΩ). If the resistance is less, suspect that C122 is shorted. Higher resistance indicates that R120 is defective or that a winding of T101 is open. Also, a leaky C124 (10 μF) could cause a low voltage at the collector terminal of Q104.

DEFECTIVE MICROPHONE

Most microcassette recorders have condenser- or capacitor-type microphones. The electret condenser microphones are found in later models (FIG. 6-27). These miniature mics are excited when a low dc voltage is applied to one side of the microphone. Sometimes an external jack is provided for an outside microphone in some models.

Electret microphone

6-27 The electret condenser microphone in this Sony M-440V microcassette recorder.

The electret condenser microphone is capacity coupled to the mic input terminals of the large IC processor (U101). The same mic terminal has 1.32 V coming from a voltage-divider network (FIG. 6-28) of R101 (560 Ω) and R102 (600 Ω). The 1.44-V source is filtered with capacitor C101. The other mic terminal goes to common ground.

These small microphones are quite sturdy except they can be damaged if sharp objects are stuck directly into the mic holes. Another low-impedance microphone can be plugged into external mic jack to see if the enclosed microphone is not functioning. These microphones should show a resistance of infinity when they are measured across the terminals and out of the circuit.

Check the bias voltage across the mic terminals when recording. No voltage might indicate that there is no supply voltage (1.32 V) or that R101 and R102 are open. Clean the contacts of the external mic jack. Remember, the external jack

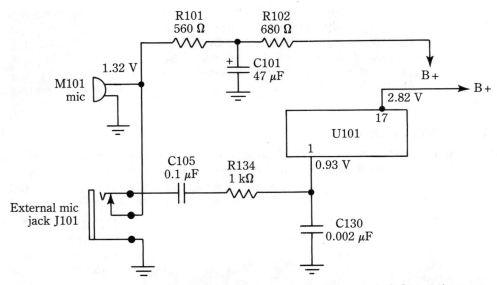

6-28 The small electret mic has + 1.32 V applied to one side of the terminals from resistor network R101 and R102.

provides a shorting circuit for the internal microphone. Short C105 with a 0.1-μF electrolytic capacitor. Any small electrolytic capacitor will determine if C105 is open. Check the voltages on pins 1, 2, 3, 4, and 17 of IC processor U101.

NO SOUND

Check the setting of the volume control. Rotate it open and listen for sound. Insert the earphone and notice if the sound is okay. If the sound is normal in earphone operation, suspect that a speaker or a dirty headphone jack is cutting the speaker out of the circuit. Clip another PM speaker across the small speaker (FIG. 6-29). Replace the speaker if it is open. Make sure that the tape heads are clean. Often, when the volume control is quickly rotated, a scratchy noise indicates that the amplifier is operating.

If you flick a screwdriver blade across the tape head with the volume wide open, you will hear a thumping noise if the amp is okay. Place the screwdriver blade next to the head ungrounded wire and listen for a hum. Check tape head for broken lead wires. With the volume control wide open, you should hear a loud rushing noise—even if the head wires are off or if the tape head is open.

If you hear a noise, suspect that the head is defective with no tape motion. Measure the continuity of the tape head. The resistance should be from 200 to 400 Ω. Very low resistance indicates that a winding is shorted within the tape head. Make sure that the tape is passing over the tape head. Clean the tape head when one channel is weak or dead, or when only one channel will not record or play back.

6-29 Clip another PM speaker across the suspected speaker terminals or wiring to determine if the speaker is open or distorted.

Next, check the transistor or the IC audio amplifier circuits. Check the supply voltage that feeds the transistor or ICs (FIG. 6-30). Shunt the input electrolytic coupling and the speaker-coupling capacitors. Clip a 100-μF 15-V electrolytic capacitor across each input and output capacitor to find the open capacitor. Test each audio transistor in the circuit with the transistor test or the diode test of a DMM.

Signal trace the audio circuits with a 1-kHz audio signal. Inject audio signal at the input terminal of the suspected IC and use a speaker as the indicator. Touch the audio signal to the output IC terminal that feeds the speaker circuit. If you can hear a low sound in the speaker, the IC might be defective. Replace the IC if the supply and terminal voltages are fairly normal with no sound output.

If U101 is in the playback mode and no sound is coming from the speaker, check the supply voltage at pin 17 (2.75 V). Measure the voltage at pins 8 and 10. Turn off the recorder and measure the resistance, which should be under 10 ohms, across pins 8 and 10 of the power output circuit. With no measurement, the speaker or external earphone jack is open. Check the resistance between 24 and 4. Measure the tape head resistance. Suspect open tape head or play/record switch contacts if the resistance measurements are very low or very high.

MUFFLED OR TINNY SOUND

A muffled or tinny sound can indicate that a voice coil is frozen against the magnet. The loose voice coil can cause a muffled sound. Substitute another PM speaker if the original is defective.

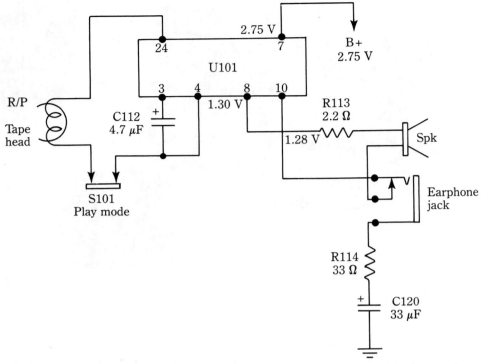

6-30 Check the supply voltage on pin 7 of U101 and check the continuity of the input and output components.

Check for weak batteries with distortion in the speaker. Doublecheck with the earphones to eliminate the speaker. If the earphones are distorted, suspect a leaky amplifier IC or leaky transistors. Clean the tape heads before tearing into the amplifier circuits. Try another microcassette. Replace it if cassette is causing the distortion.

Does the sound seem distorted when using the microphone for recording? If the recorder plays back normally, try another cassette. If the sound is distorted when a microphone is used in recording, check the microphone. Shunt the small electrolytic capacitors within mic circuits. Replace the microphone if it is defective.

Distortion can be caused by a leaky IC or leaky output transistors. Check for correct supply voltages. Signal trace the distortion with an outside audio amp. Connect the amp to the tape head and listen to a recording. If the input signal to the audio IC is normal, but distorted at the output terminal, replace the leaky IC. Check suspected leaky transistors in the same manner.

WEAK SOUND

Weak sound can be caused by weak batteries. Replace the batteries or insert an ac adapter and notice if the audio signal is still weak. Make sure that the volume control is up. Clean tape head.

A broken speaker cone might result in a weak, distorted sound. Shunt small electrolytic coupling capacitors within the audio circuits for weak sound. Suspect leaky or open power-IC or transistor-output circuits for weak signal. Check the supply voltage at the IC or output transistors.

HEAD AZIMUTH ADJUSTMENT

Suspect poor azimuth adjustment if the sound is tinny sound or if the high-frequency response is tinny. Head azimuth adjustment places the tape head horizontally with the tape path. Sometimes, weak and poor sound can be caused by poor azimuth adjustment. Try to adjust the sound with a piano or with the high notes of a recording. Insert a head azimuth test tape A411 (3 kHz). Turn the volume control up and adjust azimuth screw with an insulated tool for maximum output at the speaker (FIG. 6-31).

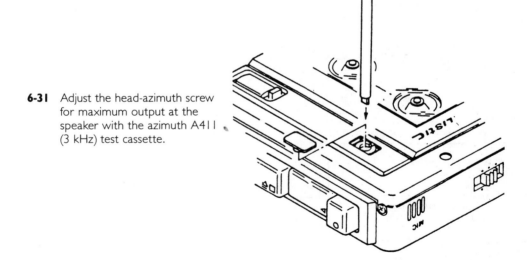

6-31 Adjust the head-azimuth screw for maximum output at the speaker with the azimuth A411 (3 kHz) test cassette.

CONCLUSION

Troubleshoot the microcassette recorder like any cassette player. Although small and surface-mounted components are crammed together to make servicing a little more difficult, you can still make many repairs. Check the most troublesome parts, such as jacks, speakers, transistors, capacitors, resistors, etc. Do not overlook dirty switches, earphone and microphone jacks.

A good tape head cleaning solves many problems with weak audio, distorted audio, or no audio, lost stereo channels, and poor recording. Clean the tape head before tackling any other sound or recording problems. Change the batteries for weak audio, poor recording, and speed problems.

A troubleshooting chart is shown in TABLE 6-1 with an exploded view of the cabinet and mechanism part locations (FIG. 6-32 and FIG. 6-33). Check the schematic diagram (FIG. 6-34) for voltage measurements of Radio Shack's microcassette recorder micro-26 (14-1043).

Table 6-1   A Troubleshooting Chart of the Various Systems Found in the Microcassette Recorder.

Tape Transport

1. No tape transport (in play mode.)

1/2

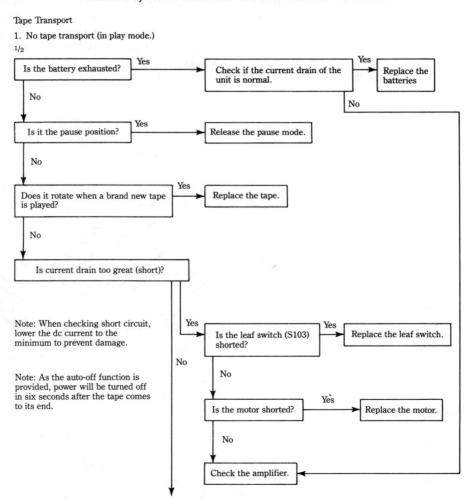

Note: When checking short circuit, lower the dc current to the minimum to prevent damage.

Note: As the auto-off function is provided, power will be turned off in six seconds after the tape comes to its end.

Table 6-1 Continued

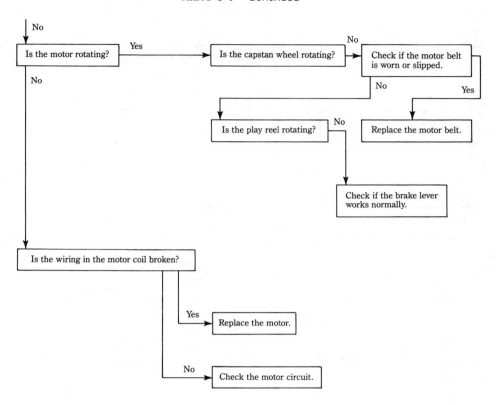

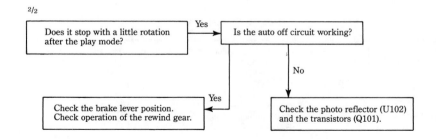

Table 6-1   Continued

2. Uneven tape transport

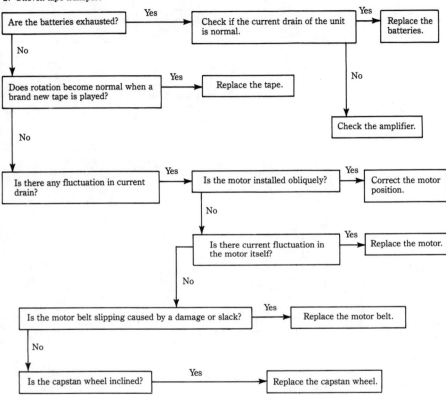

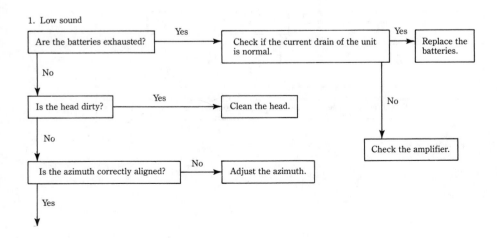

1. Low sound

Table 6-1   Continued.

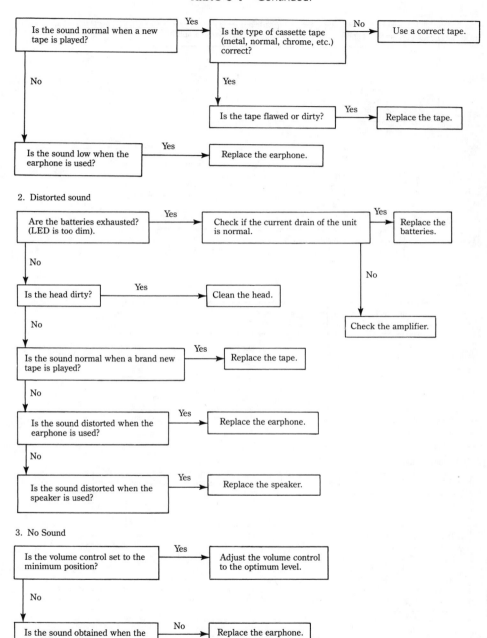

2. Distorted sound

3. No Sound

Table 6-1   Continued.

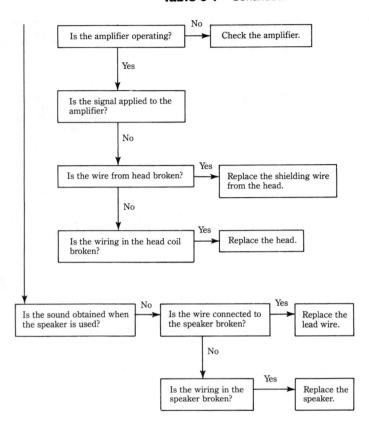

Recording

1. Distorted sound

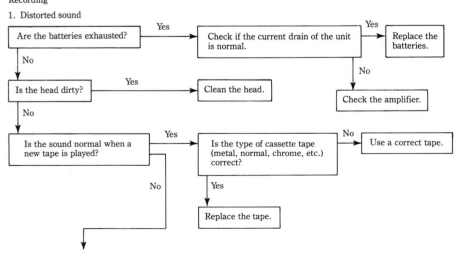

Table 6-1 Continued.

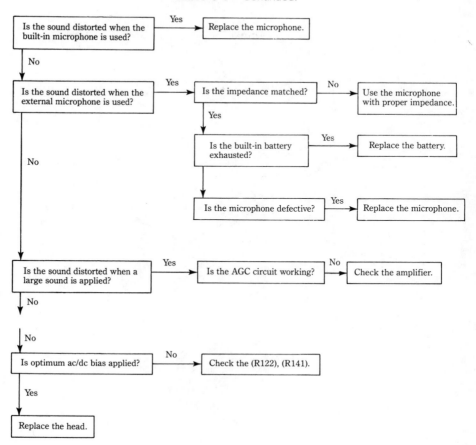

2. No recording

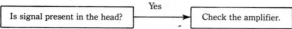

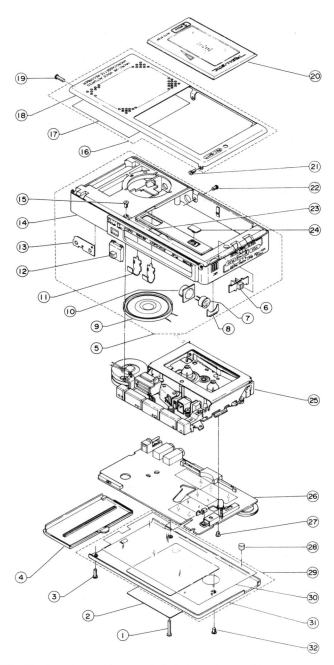

6-32 The location and the exploded view of the cabinet parts. Radio Shack

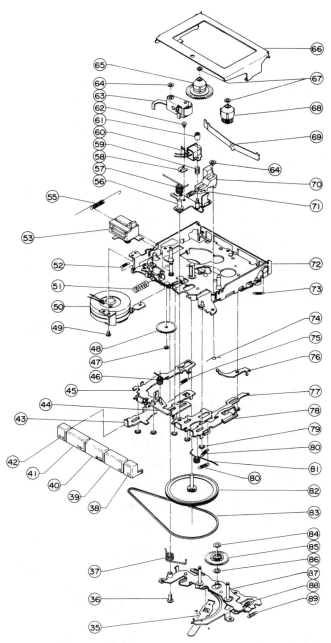

6-33 The exploded view and the parts location of the mechanism assemblies in Radio Shack's Micro-26 recorder. Radio Shack

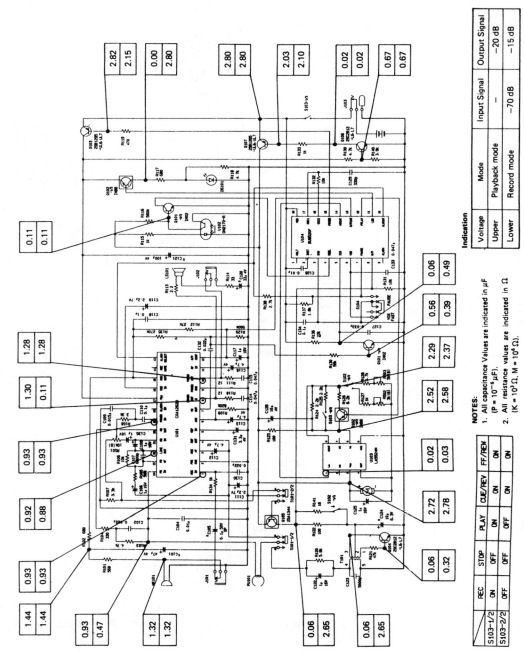

	REC	STOP	PLAY	CUE/REV	FF/REM
S103-1/2	ON	OFF	ON	ON	ON
S103-2/2	OFF	OFF	OFF	ON	ON

NOTES:
1. All capacitance Values are indicated in µF
   ($P = 10^{-6}$ µF).
2. All resistance values are indicated in Ω
   ($K = 10^3$ Ω, $M = 10^6$ Ω).

Indication

Voltage	Mode	Input Signal	Output Signal
Upper	Playback mode	–	–20 dB
Lower	Record mode	–70 dB	–15 dB

6-34 The schematic diagram of Radio Shack's 14-1043 microcassette recorder, showing voltages for recording and playing.

Chapter **7**

Servicing professional portable cassette recorders

*T*he professional portable cassette has many more features than other cassette recorders and their prices range from $99.95 up to $429.95. Most professional recorders have stereo circuits with twin microphones and speakers. Some portable stereo recorders are equipped with stereo head sets, but a few have manual circuits (FIG. 7-1).

Some of the possible features on professional recorders are two-speed recording and playback with a variable speed control, up to three heads for true monitoring, Dolby, calibrated VU meters, pitch control, and variable speech control. Headphones, stereo output jacks, and one speaker can be used as monitor. The professional cassette recorder is a rugged instrument that should stand up to years of rough treatment.

The smaller professional cassette recorders can operate from two AA batteries, and large units operate from two or three D and four AA batteries (FIG. 7-2). The recorder can have direct Quartz lock drive or disc-drive features. Some have auto stop, pre-end alarms, and manual record lever controls.

The Sony TCS-430 professional cassette recorder is compact in size and features easy one-hand operation. Another feature is adjustable tape speed in the playback mode with a playback equalizer selector for optimum playback. The TCS-430 has a built-in one-point stereo microphone (FIG. 7-3).

10 PRECAUTIONS

1. When operating on ac, always use the required voltage ac adapter.
2. When operating from the car battery, use the exact car battery cord that is recommended for the unit.

3. Always disconnect the ac power adapter from the wall outlet when the unit is not to be used for a longer period of time.

4. Keep the recorder away from radiators, heat ducts, direct sunlight, excessively dusty areas, rain. Do not drop the unit.

5. Keep strong magnets, such as speakers, metallic objects, and magnetic watches away from unit.

6. Always let the recorder warm up for a few minutes before attempting to make a critical recording.

7. Never use a cassette that plays longer than 90 minutes, except for a long continuous recording or playback.

8. Do not switch the tape operation mode frequently, which entangles or spills out the tape.

9. Keep high volume from outside sounds low when outside noise is loud and disturbing. Do not record at this time.

10. Do protect your ears by lowering the volume in extended play.

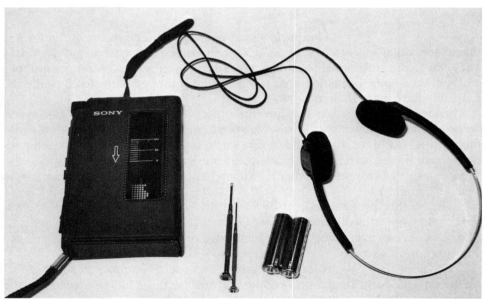

7-1   The Sony TCS-430 professional stereo cassette portable comes with a set of earphones and one speaker.

CASSETTE FEATURES

All professional recorders use the standard cassette. To remove the cassette, press the stop/eject button in the stop mode. To prevent accidental erasure of the cassette, break out the small tabs at the back. Break the tab out of side A directly to

7-2 The small professional cassette recorders operate from two AA batteries, but large units have two or three D and four AA batteries.

7-3 The TCS-430 professional cassette-corder has built-in twin stereo microphones.

the right, when looking at the back side of the cassette. Tab side B is to the left. Tape can be applied over the removed tab to record over the cassette once again.

Some professional recorders use Type 11 (CrO2) tapes or type 1V (metal) tapes. Always check with the manufacturer's literature to use the correct cassette tape. Only normal type 1 tapes are used on most recorders.

BLOCK DIAGRAM

The manually operated deluxe recorder has only one signal source of sound, but stereo units have two separate channels. Sometimes the circuit diagram lists only one channel because both stereo channels are the same. The playback and recording signals in the recorder are duplicated in the stereo units. Of course, the speed circuit's motor control, system control, and clock signals are not considered to be stereo circuits.

The block diagram is quite handy to determine where the trouble lies before looking at the large, complex stereo circuits (FIG. 7-4). Remember, only one circuit can be shown on the block diagram or schematic. The left and right channels are identical. Switching signal circuits are easily traced with the block diagram.

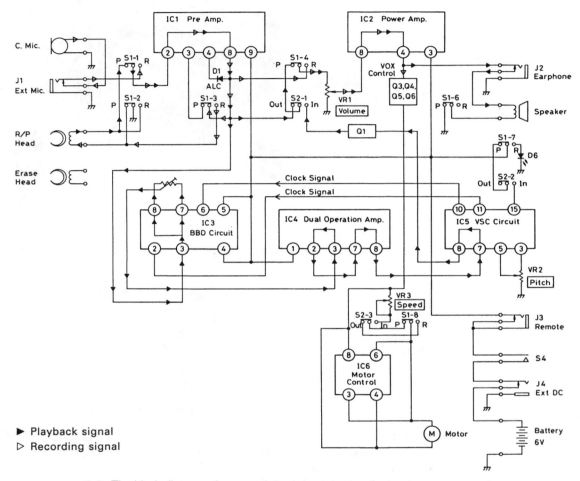

▶ Playback signal
▷ Recording signal

7-4 The block diagram of a manual signal circuit in a professional cassette recorder.

When one stereo channel is dead, distorted, or weak, the good channel can be used as a reference. The weak or dead circuit can be traced in both channels and the audio can be compared with the good channel. Also, voltage measurements can be compared to find the defective component (FIG. 7-5).

REGULAR MAINTENANCE

Clean the tape heads after every 10 hours of use to ensure optimum sound. Besides the heads, clean all tape handling surfaces: pinch rollers, capstans, and

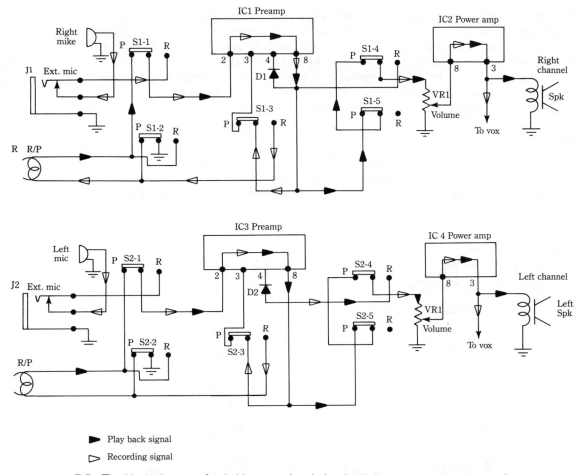

▶ Play back signal

▷ Recording signal

7-5 The block diagram of switching record and play signals in a stereo cassette recorder.

erase head with cotton swabs or cleaning sticks. You can easily clean the tape heads and pinch roller by depressing play button without a cassette in holder. This action brings the tape components out for easier excess. Remove a battery so that the recorder is not operating. Clean them with 90% denatured alcohol and a cleaning stick. After the heads and pinch rollers are cleaned, press the stop/eject button to move the heads into position so that the cassette can be inserted.

Clean the casing with a soft cloth moistened with a mild detergent solution. Do not use strong solvents such as alcohol, benzene, paint thinner, or fingernail cleaner on plastic parts. They can mar the finish or casing.

Keep all magnetized tools away from the tape heads because they might magnetize these parts. A magnetized tape head will result in increased noise, hiss, and a loss of high-frequency response. Use a standard tape-head demagnetizer to demagnetize the record/playback head.

For lubrication, use a specially formulated high-grade lubricant in the appropriate places. Lubrication is normally required only when parts bind, slow down, or make screeching noises after long periods of time. Use phono lube and lube gel for the areas to be greased. Use a drop of 3-in-1 oil or light oil on the motor and wheel bearings. The precision lubricator is handy to place grease or oil on the exact spot. Use all lubricants sparingly and avoid contact with other parts. Wipe the excess oil or grease with a cleaning stick.

DISASSEMBLY

Although the following instructions do not apply in every case, they do show how to remove the cabinet parts. Remove the battery cover and batteries. Remove the four screws that hold the rear cabinet (FIG. 7-6). Remove the screw that holds

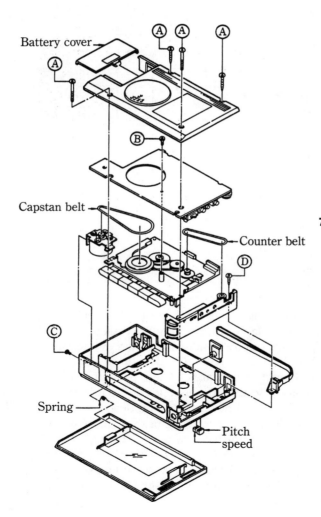

7-6 The many components to disassemble in the Radio Shack VSC-2001 variable speed control (TM) cassette recorder. Radio Shack

the amplifier PC board. Remove the capstan and the counter belts. Remove the screw that holds the motor assembly. Remove the tape mechanism. Remove the two knobs (pitch and speed). Remove the screw that holds the control panel. Remove the spring and the cassette lid. Remove four small black screws to remove the bottom panel of the Sony TCS-430 stereo cassette recorder (FIG. 7-7).

7-7 Remove four screws to remove the bottom panel of the Sony TCS-430 stereo cassette-corder.

MECHANICAL ADJUSTMENTS

Before attempting to repair or make mechanical adjustments of the recorder for slow or improper speeds, wipe all tape-contacting surfaces (R/P head, erase head, pinch roller, tape guide, and capstan). After these components are cleaned with alcohol and a cleaning stick, clean the contact surfaces of the driving parts, such as the motor pulley, flywheel, fast-forward/rewind arm assembly and all idler wheels with a piece of soft cloth soaked in alcohol (FIG. 7-8).

Of course, some drive belts, such as the capstan and the fast-forward/rewind belts can have a specially surface-treated surface and should not be cleaned with alcohol-soaked fabric. Replace any excessively greasy or oily drive belts. Clean all excess grease and oil with alcohol and a cleaning stick.

Pinch roller adjustment

Mechanical adjustments can be made with an apting gauge (0- to 500-g gauge) and a cassette torque meter. To adjust the pinch roller, clip the spring gauge to the pinch roller assembly while playing the unit. Some of these pinch rollers

7-8 Clean the contact surfaces of idler wheels, motor pulleys, belts, and flywheels with a cloth that has been soaked in alcohol.

require from 120 to 200 g of spring force. Simply hook the spring gauge to the pinch roller and pull it away from the capstan. Take a reading on the spring gauge (FIG. 7-9).

Now, measure the force at the moment when the pinch roller contacts the capstan (when the pinch roller starts rotating). If the pinch roller force is weak, adjust it by bending the spring (54) for greater tension. Replace the spring if it does not hold proper tension.

Torque adjustment

Torque adjustments can indicate if the speed of the cassette player is very slow or uneven with a torque cassette (FIG. 7-10). Make both play and fast-forward torque adjustments. The torque adjustment of a Radio Shack VCS-2001 professional cassette recorder in the play mode is 35 to 65 g/cm. Fast-forward torque should be 50 to 120 g/cm, while in the rewind mode, the torque should be from 50 to 120 g/cm (FIG. 7-11).

If the playback, fast forward, and rewind torque is very low and not in the rated area, clean the flywheel (23) and the fast-forward/rewind arm assembly (29), and check or replace the capstan belt (116) and the fast-forward/rewind belt (3). Doublecheck the seating of the torque cassette when making these tests. Do not overlook the possibility that the cassette drive motor (TABLE 7-1) might be defective.

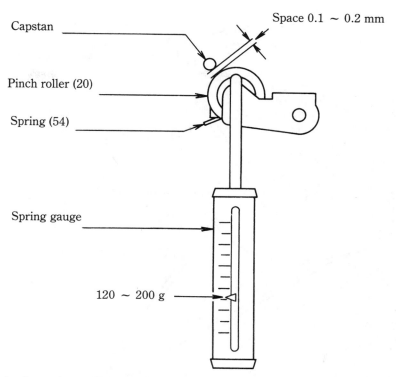

Capstan

Space 0.1 ~ 0.2 mm

Pinch roller (20)

Spring (54)

Spring gauge

120 ~ 200 g

7-9   If the fast forward, rewind, and play speeds run slow, it might be caused by improper torque of the pinch roller. Check for adequate torque pressure of the pinch-roller spring.

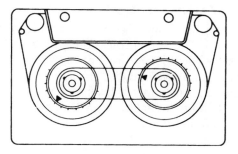

7-10   Check the fast-forward, rewind, and play torque with a torque cassette. Clean the flywheel, fast-forward/rewind belt, motor drive belt, and arm assembly for improper speeds.

ERRATIC FAST FORWARD

Often, uneven or erratic fast forward can be caused with a loose or oily drive belt. Inspect the fast-forward drive belt for oil or worn areas. Check for a bent or an out-of-line fast-forward arm assembly. Notice how the different idlers and pulleys engage. A dry bearing of an idler wheel or turntable reel can cause erratic fast forward (FIG. 7-12). Doublecheck the fast-forward speed with a torque cassette.

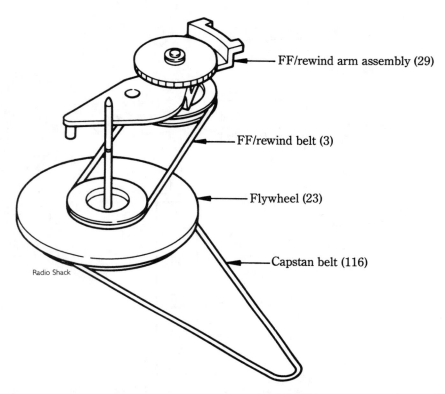

FF/rewind arm assembly (29)

FF/rewind belt (3)

Flywheel (23)

Capstan belt (116)

Radio Shack

7-11   The correct fast-forward torque for Radio Shack's VSC-2001 cassette recorder was 50 to 120 g/cm on the torque cassette.

ONLY ONE SPEED

Some professional cassette recorders have two different speeds or a variable speed control. The Sony professional recorder has a variable speed control located outside for the operator to control (FIG. 7-13). The variable-speed control is located in the motor drive circuits. The actual tape speed is 4.8 cm/sec (1.78 lps) and adjustable 10 to 15% in playback mode.

Within Radio Shack's VSC-2001 cassette recorder, a 2-kΩ variable-speed control is found within the play or record circuits of IC6. Several factory preset controls are found in this speed circuit (VR-9, VR-10, and VR-11). The speed of the cassette motor is controlled by pins 3 and 4 of IC6 (FIG. 7-14). The cassette-motor 6-V supply is fed through external dc jack 4, leaf switch S4, and remote jack J3. Variable speed is only adjustable in playback mode, not during recording.

Take critical voltage measurements on terminals 3 and 4, and 6 and 8. Clean switches S2-B and S1-B with alcohol or cleaning fluid. Erratic speed can be caused by a worn or dirty speed control (VR3, 2 kΩ). Spray cleaning fluid into the controls. Suspect that IC6 is defective if a variable voltage is going into pins 6 and 8 and no change is at pins 3 and 4. Check all resistors in the speed-control circuits for increased values.

Chart 7-1. Preliminary Troubleshooting Chart

Symptom	Cause	Remedy
Cassette cannot be inserted	Cassette inserted improperly.	Check cassette. Check for foreign material inside. Notice if play or record button depressed.
Record button cannot be depressed.	No cassette loaded. Cassette tab removed.	Reload cassette. Check for small tab removed at rear of cassette.
Playback button cannot be locked in.	Check for complete wound tape.	If tape is completely wound toward arrow direction, rewind tape with rewind button.
Tapes does not move.	Incorrect battery polarity. Weak batteries. No ac.	Batteries in backwards. Test and replace if below 1.2 V. Power adapter not connected.
No sound from speaker	Are headphones plugged in? Check location of volume control.	Remove headphone plug. Volume turned down.
Tape speed excessively fast	Check setting of speed control.	Readjust speed control.
Weak or distorted sound	Weak batteries. Dirty heads.	Test or replace. Clean with alcohol and cleaning stick.
Poor recording	Weak batteries. Dirty stereo heads.	Test or replace. Clean with alcohol and cleaning stick.
Poor erasing	Improper connections. Dirty erase head.	Check all cord and wire connections. Clean erase head.

7-12 Erratic fast forward might be caused by a loose or oily fast-forward belt. Check and clean all fast-forward drive surfaces. Replace worn or old belts.

7-13 The variable speed control is located on the outside of this Sony professional portable recorder.

UNEVEN PRESSURE ROLLER

A worn or damaged pressure roller can cause the tape to spill out and feed unevenly. The pinch roller is quite small in the portable cassettes, although not as small as the roller in microcassette players. Check the pinch roller for worn spots. Determine if enough pressure is applied to the pinch-roller assembly with a small pressure gauge, if it is handy. Bend the tension spring for more pressure if the gauge measurement is low. Replace the entire pinch-roller assembly if it is defective (FIG. 7-15). Excessive wow can be caused by a dirty pinch roller.

NO TAPE MOTION

Check the batteries and replace them if they register less than 1.2 V. Clean the on/off leaf switch. The leaf switch is on in all functions (FIG. 7-16). This switch is in series with the external dc jack (J4), batteries, and remote jack (J3). Sometimes, these leaf-switch contacts become dirty and bent together.

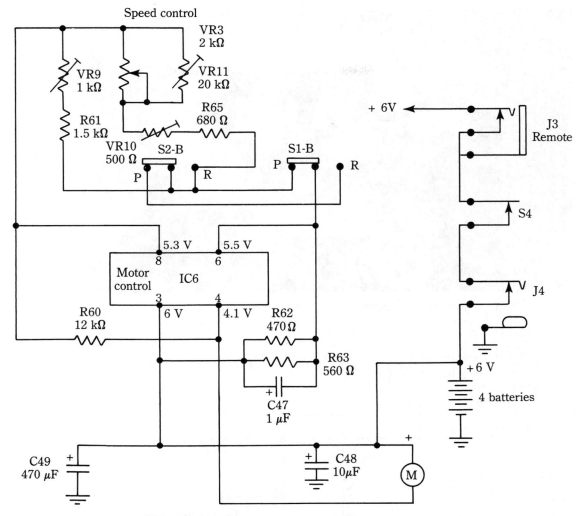

7-14 The variable speed control in the cassette motor-driver circuits.

If you find a voltage at the power-output IC and on both sides of the switch, suspect a defective motor circuit or motor when the motor will not rotate. Check for voltage across the motor terminals. If voltage is present and the motor doesn't rotate, replace the drive motor.

Sometimes you can hear the motor run, but nothing happens in the play, record, fast-forward, or rewind modes. Remove the bottom cover and notice if the capstan/flywheel is rotating. Check for a missing or broken belt. Erratic operation can be caused by an oily or overly large motor drive belt. Wipe the flywheel surface and belt with alcohol and a cleaning cloth. Replace the capstan belt if it is loose or excessively greasy.

7-15   Replace entire pinch-roller assembly if it is worn, bent, or defective.

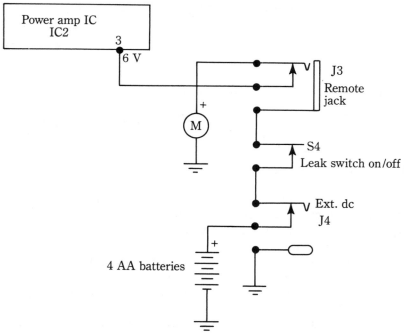

7-16   IC2 and the cassette motor receive supply voltage through the remote control, leaf switch (S4), and external dc jack (J4).

EXCESSIVE WOW

If wow is excessive, clean the pinch roller and capstan drive. Check the pinch-roller pressure with a pinch-roller gauge. Replace the pinch roller if it is worn or deformed. Check all drive belts and clean them. A defective drive motor might produce uneven speeds. Check the motor drive voltage from the IC or transistor voltage-regulator circuits. If the voltage is constant and the motor operation is erratic, replace the defective motor (FIG. 7-17).

7-17 A close-up view of the small motor in Sony's TCS-430 stereo cassette recorder.

Small dc motor

NO SPEED CONTROL

If the speed control doesn't work, check the setting of the speed control. Excessive tape speed can be caused by an improper setting of speed control. Clean slide switch S2 and replace it, if it is defective. Check the speed control for open or dirty contacts. The total resistance across control is 2 kΩ. Measure the voltages on all terminals of motor control IC6 (FIG. 7-18). Check the components around the IC. Replace IC6 if the voltages are fairly normal without the speed control.

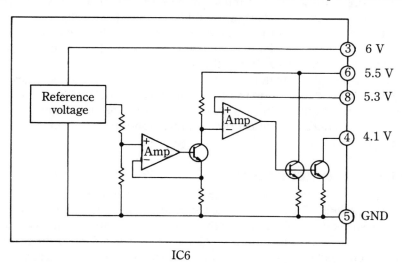

IC6

7-18 The internal stages of motor control IC6 with the operating pin-terminal voltages.

NO LEFT MIC CHANNEL OPERATION

Stereo is used in most professional cassette recorders. Dual condenser microphones are contained in the left and right stereo channels of Sony's stereo cassette recorder (FIG. 7-19). Here, stereo headphones are used in playback and one speaker is used as a monitor. The sound recorded can be heard through the headphones. You can adjust the monitor volume with the volume control.

7-19 Two electric mics in a stereocassette player.

The condenser or electret microphone has a bias voltage applied to the ungrounded side of the mic terminal. With external mic jacks, the built-in microphones are placed in series with the external mic jack (J1). Suspect a defective mic when the right channel sounds normal and only the left channel is dead. Also, check the switch contacts on S1-1.

Try another external electret microphone in the external jack to determine if the internal mic and circuits are defective. If the external jack mic operates, take voltage measurements across the microphone terminals (1 to 1.5 V). Suspect R4, R5, and C3 if you find no voltage at the mic terminals (FIG. 7-20).

If both the external and internal microphones are dead, check switch S1-1 and IC1. Does the left channel play a cassette normally? If so, the R/P switch or mic wires might be defective. When the left channel is completely dead on play and record, suspect that preamp IC1 is bad. Measure the supply voltages at pin 9 (5.1 V). Check all voltage terminals of preamp IC1 (FIG. 7-21). If the preamp stages are transistors, check the voltages in the microphone input and preamp circuits. Remember, two different identical preamp microphone circuits are found in the stereo recorder.

POOR LEFT-CHANNEL SOUND

If the left-channel sound is poor, clean the tape heads. Is the poor left-channel sound in the playback or record mode? Start at the preamp IC and work toward the monitor speaker or earphones of the stereo left channel if the poor sound is found in both the recording and playback modes. Try a new recording cassette to see if the left channel still has poor sound. Weak sound can occur in any stage, but distorted sound is often found in the audio output circuits.

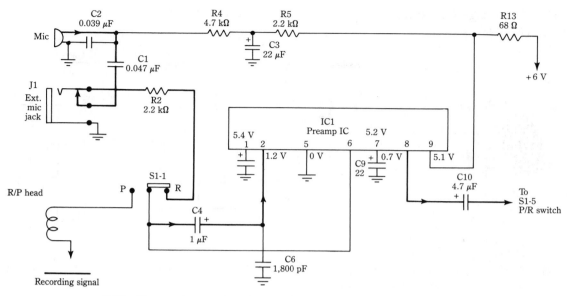

7-20 The condenser-microphone recording-signal circuit of preamp IC1.

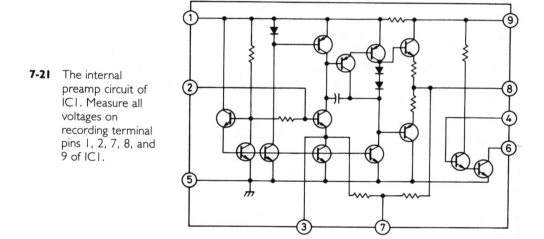

7-21 The internal preamp circuit of IC1. Measure all voltages on recording terminal pins 1, 2, 7, 8, and 9 of IC1.

Weak sound is more difficult to locate than a dead symptom. Check small electrolytic coupling capacitors, transistors, bias resistors, and tape heads. A low supply voltage can create weak sound in either transistor or IC circuits. Test for a change in resistance or capacitance when connected directly to the IC terminals.

Distorted sound can result from a dirty or worn tape head. Bias resistors and leaky output transistors or ICs can cause low distortion. Shorted or leaky resistors and electrolytic coupling capacitors can cause distortion. Insert a 1- or 3-kHz cassette in the recorder and signal trace the audio with an outside amplifier.

Compare the signal of both audio stereo channels to locate the defective component (FIG. 7-22).

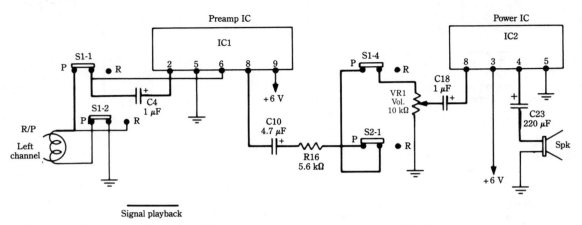

Signal playback

7-22 Signal trace the audio in the preamp (IC1) and power IC output (IC2) for weak or distorted sound and compare them with the good stereo channel.

POOR HIGH-FREQUENCY RESPONSE

When music sounds high-pitched and tinny, suspect a defective R/P head, improper azimuth adjustment, or a broken speaker cone. Inspect the tape head for worn spots on the front area where the tape passes over. After you clean the tape head, if the R/P head looks normal, check the azimuth adjustment. Insert a 3- or 6.3-kHz test tape (MTT-113N) and adjust the azimuth screw for maximum volume at the speaker. This screw is located beside the tape head and it moves the head horizontally with the tape (FIG. 7-23). Replace the tape head if it is defective.

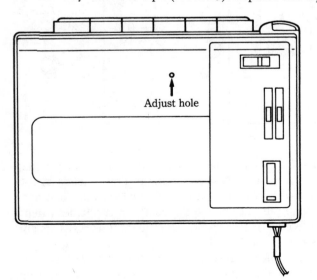

Adjust hole

7-23 The location of the tape head azimuth screw in the Radio Shack VSC-2001 recorder.

Check the speaker for a tinny or muffled sound. Of course, this sound will be constant if the speaker cone is frozen. The cone is warped and lays against the center pole piece, which creates the tinny, muffled sound. Replace the speaker if it is defective.

NO SPEAKER/MONITOR

In many of the small professional cassette recorders, one speaker is used as a monitor (FIG. 7-24). When the stereo headphones are inserted, the speaker is automatically removed from the circuit. When recording, the sound can be heard through the headphones. Keep the headphone volume low or a howling sound (feedback) might occur. When you take off the headphones during a recording, lower headphone volume control, then take them off. Otherwise, a howling noise might occur.

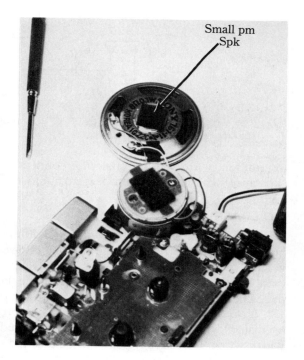

7-23 The location of the tape head azimuth screw in the Radio Shack VSC-2001 recorder.

When the headphones are working, but the monitor speaker isn't, suspect that a jack contact is poor, speaker wires are broken, or a speaker is defective. Clip another small PM speaker across the built-in speaker terminals. If the outside speaker works, replace the defective speaker. If you hear no sound, check the speaker-wire terminals. Inspect the earphone shorting jack for possible damage or broken terminal wires (FIG. 7-25). Notice that R23 (33 Ω) is in the circuit to lower the headphone volume, compared to the monitor speaker.

Suspect that C23 (220 μF) is defective if the player has no speaker or earphone reception. These speaker-coupling capacitors can open up or become

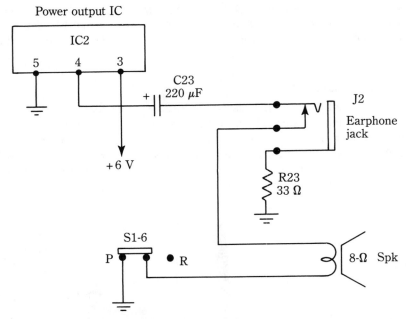

7-25   The audio signal is taken from pin 4 of IC2 through C23, J2, and the speaker with the speaker return wire through switch S1-6.

intermittent with poor internal wire connections. A poor internal terminal wire can cause weak, intermittent, or dead audio.

NO ERASE

If you find jumbled recordings on several different cassettes, suspect that the tape head is erasing poorly. Clean the erase head. Some erase heads are excited by a dc voltage or a PM magnet. The erase head is excited with the bias oscillator. The erase head might be open or shorted. Replace the erase head if it is open or defective. Check the erase-head continuity with the ohmmeter.

If you find both recording and erasing problems, suspect a defective bias-oscillator circuit. The waveform can be checked at the high side of the R/P head with a scope. If a scope is not handy, check the oscillator transistor voltage, T-300 windings, and resistance changes in the oscillator circuit (FIG. 7-26). Some recorders have a BBD clock and ramp generator in a VSC (variable sound circuit) chip, which feeds a BBD IC, which excites the erase and R/P tape heads.

In large professional cassette recorders, you can find a separate erase head. With the small units, the erase head can be located inside the R/P head assembly. Peek at the front side of the R/P head to see if you locate another gap area. The stereo R/P and erase head has three gap areas. If the erase head is located inside the stereo R/P head, you will find six terminal hook-up connections or wires. Remember, in a stereo record/playback head, the record and playback modes use the same head for both functions. Thus, it contains two different gap areas.

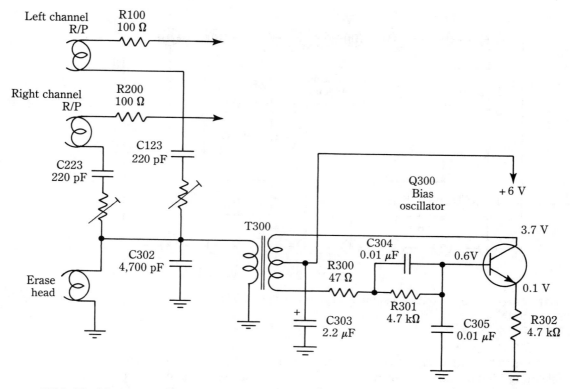

7-26 The bias oscillator might excite each stereo tape head and erase head in larger recorders.

POOR RECORDING

Poor or erratic recording can result from a defective erase head that does not completely erase the previous recording. Clean the R/P and erase head with alcohol. Sometimes one of these small gaps can become packed with tape oxide dust, which results in no recording. Try another cassette.

Check the bias oscillator for erratic or intermittent operation with poor recording symptoms. Poor oscillator transformer connections can produce intermittent recording. Weak recording can be caused by dirty record head or a low supply voltage to the bias oscillator transistor. Check for poor or dirty record switch contacts. Clean them with cleaning fluid. If both the record and play is weak or distorted, check the audio output and preamp circuits.

NO PITCH CONTROL

The pitch slide switch (S2) is located in terminal pin 15 of IC5. S2-2 switches in a B+ 6-V supply voltage for IC5 or simply turns on the pitch-control circuits (FIG. 7-27). The pitch-control range is controlled by VR2 (5 kΩ) at pins 3 and 5. Check slide switch S2-2 if it has no pitch control. Measure the voltage (6 V) at pin 15. Clean S2-2 if the pitch control is intermittent or dead.

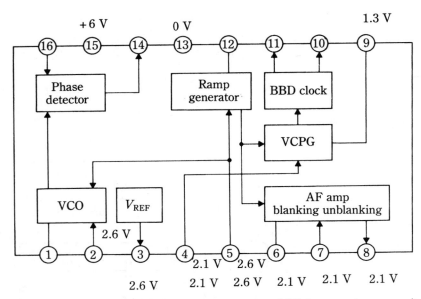

7-27 Check the supply voltage and the voltage on all pins of IC5 for no or improper pitch control.

If pitch control VR2 is open, the pitch control might be intermittent. Spray cleaning fluid inside the control area if VR2 is dirty. Measure the resistance across the control to determine if it is open without pitch control. Do not overlook the possibility that the variable pitch control, IC5, IC3, IC4, or the corresponding components might be defective.

NO VU MOVEMENT

Some manual professional portable cassette recorders have a single VU meter, but the stereo recorders generally have two separate VU meters. The VU meter measures the recording amplitude or how loud the recorder is recording the audio. When one channel becomes weak, the VU meter in that channel indicates that the recording response is weak. The VU meter can be used to signal trace the audio circuits.

The VU meter can indicate if the sound is normal after the Dolby amp circuits; some are located before the line-output terminals. VU or signal meters can be checked with the ohmmeter. Remove one lead and set the ohmmeter to the RX1 scale. A good meter will read to under 300 Ω. Notice the rise of the VU meter when the ohmmeter is connected across it. Replace the meter if it is open. Check diodes D501 and D502 (FIG. 7-28).

The pointer on the VU meter can stick or rub on the dial plate. Sometimes the dial cardboard will warp and bend upwards, which stops the movement of VU meter hand. Repair the meter by disconnecting the terminal wires and mounting screws. Remove the cabinet and the front meter glass. Reglue the dial plate and

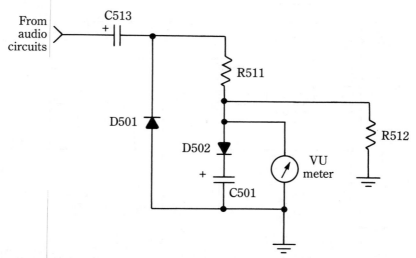

7-28 The VU-meter circuit attaches to the audio signal and is rectified by D501 and D502.

reassemble the meter. Turn the small set screw at the bottom to zero. Check for audio up to the VU meter if you suspect that it is the problem.

TROUBLESHOOTING

If possible, pick up a schematic diagram of the recorder for easy troubleshooting (FIG. 7-29). A service manual of the exact cassette recorder has the PC board wiring layout (FIG. 7-30) and wiring diagrams to help you to locate defective components (FIG. 7-31).

Besides a layout of electrical and electronic components, the exploded view of the tape mechanism helps you to see where the various parts tie together (FIG. 7-32). Sometimes, the exploded view of the cabinet parts show how they are mounted and placed together (FIG. 7-33). For additional troubleshooting symptoms, follow TABLE 6-2.

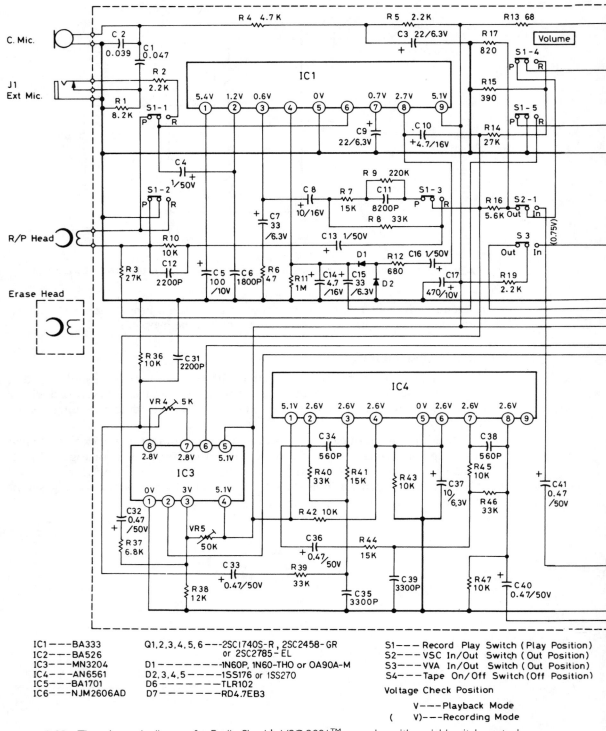

7-29 The schematic diagram for Radio Shack's VSC-2001™ recorder with variable pitch control.

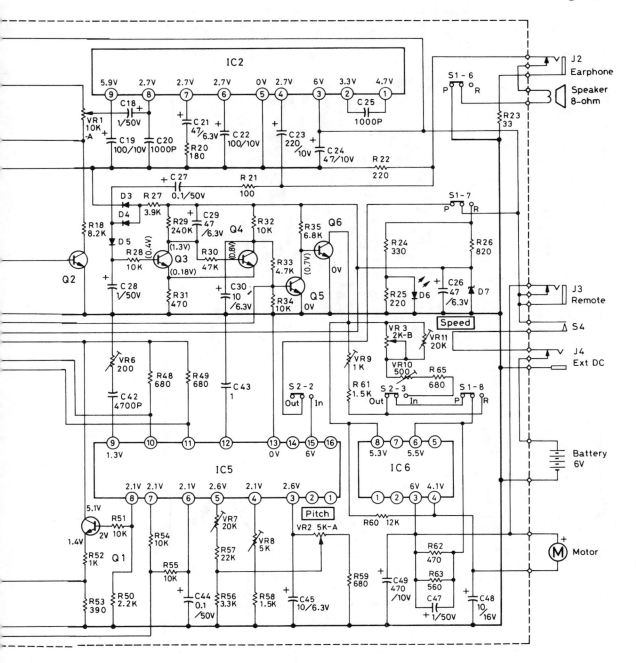

Notes:
1. All resistance values are in Ω. k = 1000Ω M = 1000KΩ
2. All capacitance values are in μF. P = 10^{-6} μF
3. All resistors are 1/6 watt, unless otherwise specified.
4. Voltages measured from point indicator to chassis ground with SSVM at line volume control minimum and no signal.

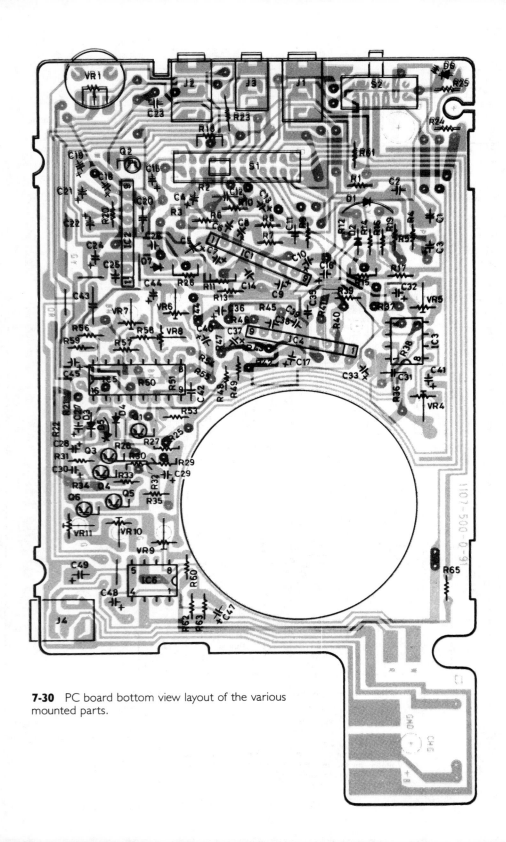

7-30 PC board bottom view layout of the various mounted parts.

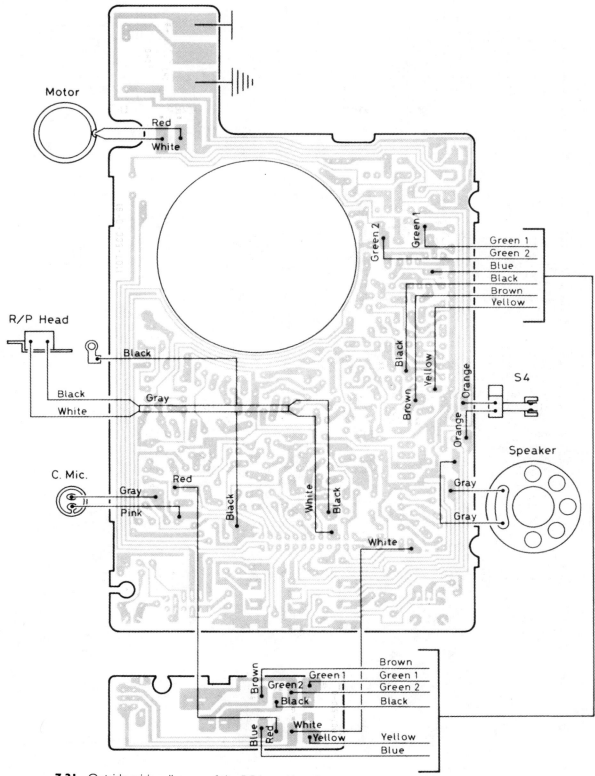

7-31 Outside wiring diagram of the PC board in a Radio Shack VSC-2001 recorder. Radio Shack

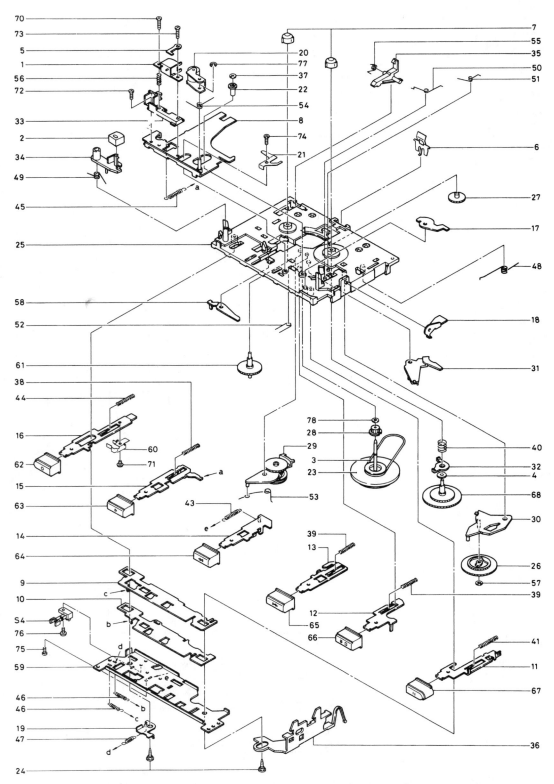

7-32 An exploded view of the tape mechanism showing connecting parts.

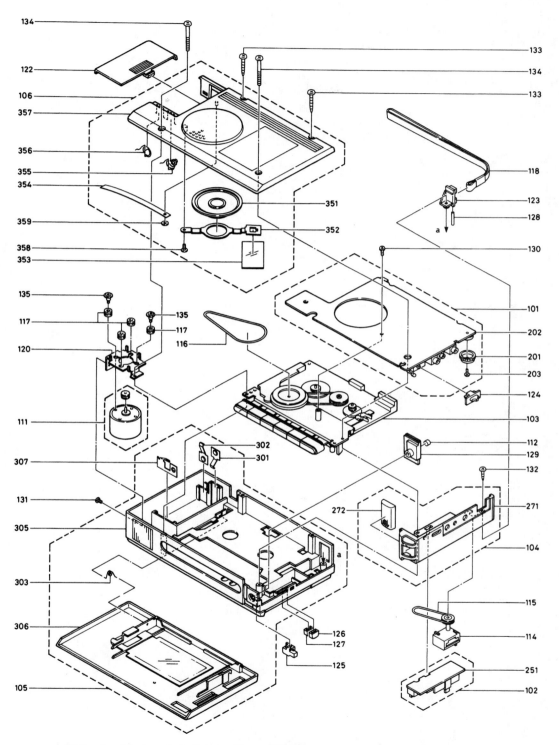

7-33 A cabinet exploded view of how different components are mounted and listed.

Chapter **8**

Troubleshooting car stereo cassette players

Today's average auto stereo cassette player has digital tuning with 12 FM and 6 AM presets, seek/scan tuning, and at least 6 W output (FIG. 8-1). The deluxe unit might have quartz electronic tuning, 18 or more presets, an LCD display, auto reverse, from 7 to 20 W output, and 4 to 6 PM speakers. High-power car players have 40- to 160-W stereo amplifiers (FIG. 8-2).

Surface-mounted components can be found in the electronic digital tuning, clock, system control, and AM/FM IF circuits (FIG. 8-3). These surface-mounted components can be mounted on the PC board side, and the other ICs and transistors appear on the other side. An intermittent surface-mounted component can be difficult to locate. High-powered ICs and transistors are used in the audio circuits.

BLOWS FUSES

Suspect a leaky power-output IC or transistor, pinched connecting wires, or a shorted filter capacitor when the fuse keeps opening. Inspect the hookup harness for bare wires or improper connections (FIG. 8-4). Make sure that the American radio has a negative ground and a positive ground to foreign autos. The auto-battery polarity might be switched or battery might be charged backwards.

Go directly to the audio output IC or transistor if you see smoke, burned wires, or scorched PC board wiring. Often, the power IC uses the car radio chassis as a heatsink (FIG. 8-5). Check for leakage from each terminal to ground. With burned PC board wiring, the defective output IC and transistor can be quickly located. Besides replacing all defective components, all wiring and PC board repairs must be made.

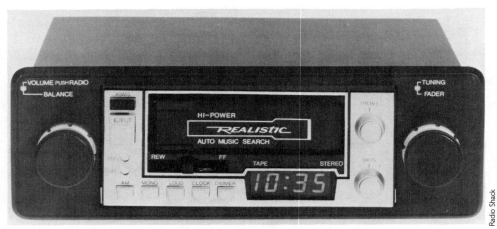

8-1 The high-power auto cassette player might have auto music search and digital tuning.

8-2 The auto radio-cassette player might have a high-powered amplifier that could output up to 160 W.

In the early auto radio chassis, power output transistors were used. Today, one large power IC or two different channel ICs provide audio to the speakers. Suspect the output dual-channel IC when both channels are dead, distorted, weak, or intermittent. Signal trace the audio up to the input terminals of the power IC.

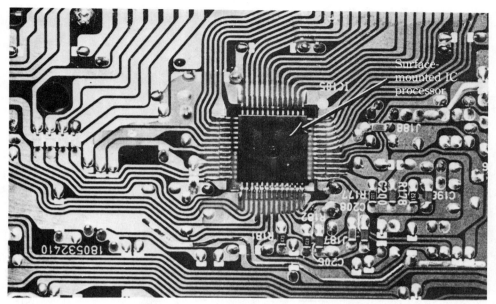

8-3 Surface-mounted components might consist of IC processors in the digital and clock circuits of the latest auto radios.

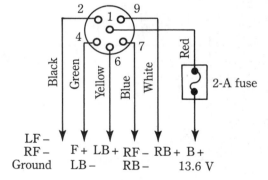

8-4 Inspect the harness for bare or burned wires.

The balance, tone, and volume controls are found in the input circuits of each power output channel (FIG. 8-6). The left channel input audio is found at pin 13, and the right channel IC pin is 2. The left output signal is taken from pin 8 through coupling capacitor C51, and the right output from pin 7 through electrolytic capacitor C52. When either C49, C48, C51, or C52 dries up, the output signal is weak. If either capacitor opens, the sound might die at the speaker.

PILOT LAMP REPLACEMENT

In most cases, you must remove the auto radio from the car to replace dial light bulbs. Regular 12- or 14-V light bulbs plug into a bayonet socket and should be

8-5 A leaky or shorted power-output IC might cause the fuse to blow or open.

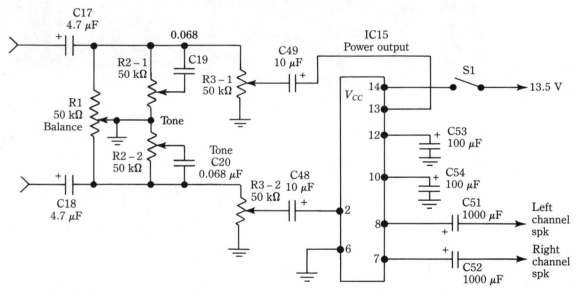

8-6 One large power IC might include both left and right stereo output circuits, with balance control.

checked when the covers are off the radio. If the bulb still lights, but the glass is black, replace it because they will die in a few months (FIG. 8-7). Light bulbs with long leads can be replaced by cutting off the leads and soldering them into place. A drop of glue on end of bulb helps hold bulb into position. These bulbs can be purchased at most auto or electronic stores.

LEDs are often used as light indicators in the latest auto cassette players. LEDs are used as pilot lights, normal and reverse direction indicators, switching lights, and fast-forward and rewind lights (FIG. 8-8). In larger cassette players, Dolby, AMSS, and stereo LED are used. Usually, LEDs will last the life of the cassette player. Pilot lights are fairly easy to replace after the unit has been removed from the auto.

8-7 Many different light bulbs and LEDs function as indicators in this auto cassette player.

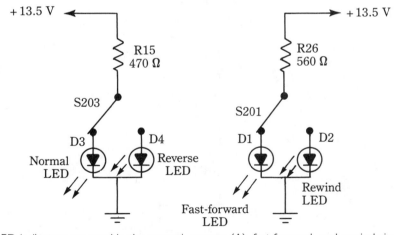

8-8 LED indicators are used in the normal, reverse (A), fast-forward, and rewind circuits (B).

ERRATIC SPEED

Intermittent tape speed can result from an oily belt, a dry capstan, or from a defective drive idler, speed-control circuit, or motor. Check the voltage at the motor to isolate a defective speed circuit. Start with a new cassette. Clean all dry surfaces with cleaning stick and alcohol.

If you clean and replace the loose drive belts and the speed is still erratic, suspect that the motor is defective (FIG. 8-9). Monitor the motor voltage at all times. Sometimes tapping the end bell of motor will make it change speeds. When the voltage is not constant at the motor terminals, suspect that a speed circuit is defective or that a radio/tape switch is dirty (FIG. 8-10). If the motor is working directly from the battery source without a speed-regulator circuit and is intermittent, install a new motor.

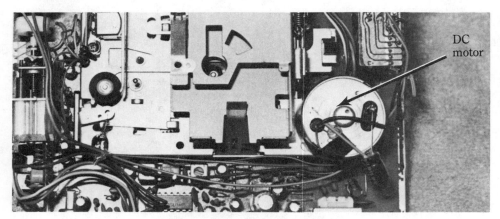

8-9 A defective motor might cause erratic or fast or slow speeds.

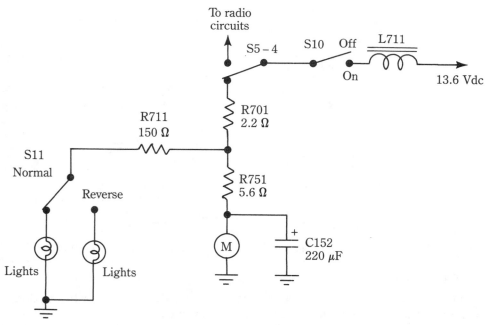

8-10 Improper voltage at the motor terminals might be caused by a defective speed circuit, isolation diode, resistor, or dirty radio/tape switch terminals.

HIGH SPEEDS

Most higher-than-normal speed operation is caused by either a motor drive belt riding on the rim of the motor pulley, a defective motor regulator circuit, or a defective motor. A leaky IC or transistor motor regulator circuit can cause high speeds. Check and monitor the voltage at the motor circuits (FIG. 8-11). Readjust the speed control in regulator circuits.

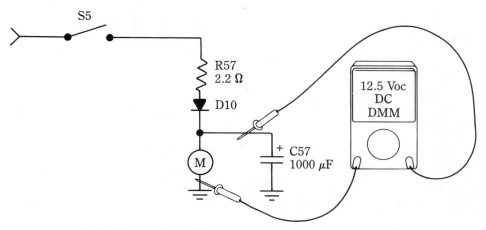

8-11 Monitor the voltage across the motor terminals if the motor is intermittent or rotating at a faster speed.

In some regulator circuits, two different adjustments are provided for normal and reverse rotation. Insert a 3-kHz test tape and connect a frequency counter at the speaker output. Adjust both the reverse and normal controls for a 3,000-Hz reading on the frequency counter.

The motor speed regulation circuit can be a transistor or an IC. The regulator circuit provides accurate voltage and keeps the motor operating at a constant speed. If the transistor or IC becomes open or leaky, the speed can increase or lower, according to the dc voltage that is applied to the motor (FIG. 8-12). Test each transistor within the motor regulator circuit. Sometimes, these intermittent transistors will only underload. If in doubt, replace the transistor or IC. Check each diode and resistor within the speed-regulator circuit.

If the motor circuits and voltage applied are normal, replace the defective motor. The speed in the defective motor can run high or low. Often, the speed is low, sometimes caused by dry bearings. A squirt of light oil at each motor bearing can let the motor resume normal speed. A screeching or squealing motor indicates that the bearings are dry. Check the cost of a new motor before removing a defective motor.

WORKS ON RADIO/NO TAPE ACTION

Listen to the motor when operating in fast forward or play. If the spindles do not rotate or if you can't hear the motor, suspect that the voltage is not getting to the

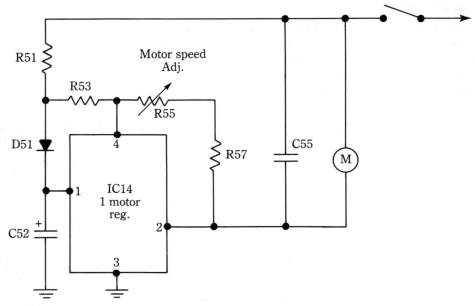

8-12   The defective motor-regulator circuit might cause the motor to rotate at high speeds.

motor. Measure the voltage at the motor terminals. Suspect a dirty off/on or radio tape switch. Clean the switches. Check the dc voltage after each switch. If voltage is on both terminals, but none is at the motor, suspect that the isolation resistor is open (FIG. 8-13).

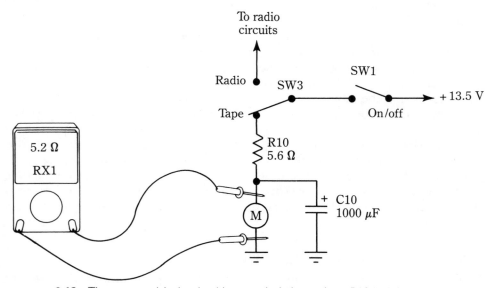

8-13   The motor might be dead because isolation resistor R10 is defective.

WILL NOT LOAD

The cassette can load edgewise or full-on in the auto radio/cassette player. When the cassette is pushed in, a lever is engaged and the cassette platform snaps next to the capstan/pinch roller assembly. A missing or broken tension spring cannot let the holder load properly. A bent cassette or levers prevents loading (FIG. 8-14). Inspect all levers and bent areas. Apply light grease to the sliding areas.

8-14 Suspect that the carriage or levers are bent if the cassette will not load in a Clarion 9100 RT auto cassette player.

Look for foreign objects inside the cassette loading area. Sometimes gum wrappers or cigarette butts jam the loading area. If cassette player is near a cigarette tray, remove the player and clean out any cigarette ashes, which can cause binding or sliding metal parts to malfunction.

JAMMED TAPE

If the cassette will not unload, it might be caused by a malfunction in the loading platform or a tape spill out. The tape might be wrapped tightly around the capstan and pinch-roller assembly. Sometimes, the capstan/flywheel can be reversed by hand to release the tape tension. Often, the tape must be cut loose because the cassette metal housing prevents removing excess tape and the cassette easily.

After removing the excess tape and the cassette, clean the capstan and the pinch roller. Clean them thoroughly with alcohol, a cleaning stick, and a cloth. Wipe off all excess tape oxide within the loading area. Clean the tape heads while the unit is open (FIG. 8-15). See if the cassette will load easily.

ERRATIC OR INTERMITTENT AUDIO

In auto cassette players with dual-capstan/flywheel and dual heads, switching of head contacts can cause erratic or intermittent operation. Spray cleaning fluid into

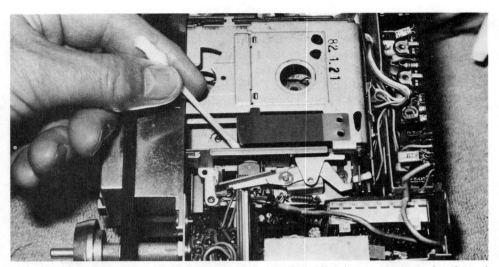

8-15 After removing a jammed cassette from the holder, clean the capstan, tape heads, and pinch roller with a cleaning stick and alcohol.

the head switching contacts. Check the small wire head leads for broken wires at the head terminals. When these heads are moved, the wires could break off (FIG. 8-16). Sometimes, the wire breaks inside the insulation, but still appears normal. Lightly pull on the small wires with a pair of long-nose pliers to uncover the broken connection.

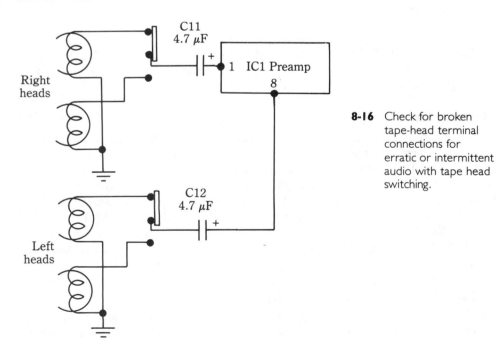

8-16 Check for broken tape-head terminal connections for erratic or intermittent audio with tape head switching.

Moving the head or switching assembly will eventually break the head wires. Pull back the insulation and tin the flexible wire before soldering it to the head terminal. Be careful not to leave the iron on the terminal so long that it dislodges solid terminal from the back of the head.

AUTO STEREO CHANNELS

In early solid-state car cassette players, transistors were used throughout the audio circuits. Now, very few transistors are used because ICs took over. The simplest IC audio channels consist of one dual IC preamp and a dual output IC. The tape head is directly coupled to the preamp IC through a small electrolytic capacitor (FIG. 8-17).

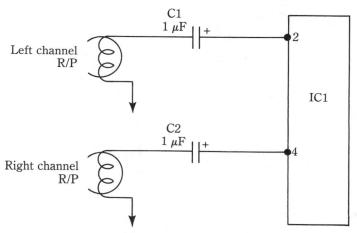

8-17 The tape head is directly coupled through a small electrolytic coupling capacitor to the IC.

When the preamp IC circuits appear to be defective, take critical voltage measurements on IC1, check the continuity of the tape heads, and signal trace the audio output at pins 3 and 6. Do not overlook the voltage regulator circuit when a low or improper voltage is found at pin 4. Some auto cassette players have a separate voltage regulator that operates directly from the dc battery source (FIG. 8-18).

First, take voltage measurements on all terminals of Q9. Low or no voltage at the emitter terminal might indicate that a transistor is leaky or open. A low voltage output at the emitter might result from a leaky zener diode (D10). Check both the transistors and the diodes with the diode test of the DMM. If D10 becomes leaky, R40 might change in value or burn.

Usually, the tone, volume, and balance control are located in the input stereo circuits of the audio output IC circuits. Signal trace the audio at the volume control and into pins 4 and 2 of the IC output. Suspect a defective IC or circuit components if the output signal is improper at pins 7 and 8 (FIG. 8-19).

Troubleshoot the output IC with critical voltage measurements. Compare the good channel with the defective voltages. The supply voltage is at pins 1 and 14.

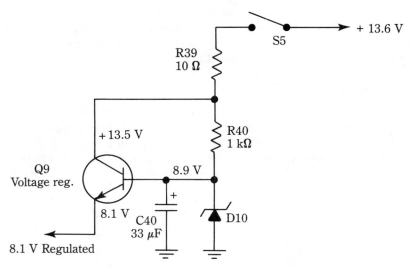

8-18 Do not overlook the possibility that the preamp voltage transistor regulator might be dead if you measure an improper voltage at the preamp circuits.

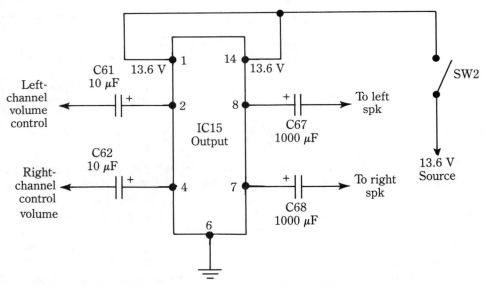

8-19 Suspect that IC15 or that the supply voltage at pin 14 is bad if no audio signals are at pins 7 and 8.

If this voltage is low, check the dc source and the large filter capacitor. An open choke or fuse can prevent the voltage from arriving at pins 1 and 14 (FIG. 8-20).

Larger auto cassette players with higher wattages have separate power output ICs or transistors. These power circuits are serviced in the same way as the dual units. However, in this case each channel component is separate and when the

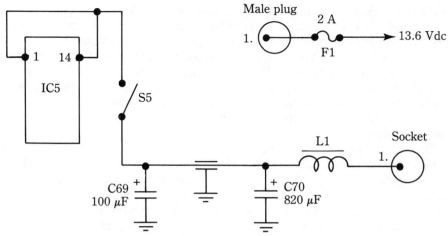

8-20 An open fuse (FI) or choke coil LI might prevent the dc voltage from being applied to switch S5.

dual IC becomes defective, the whole unit must be replaced—even if only one channel is dead or distorted. The left channel input audio is at pin 10 and the output is at pin 2 (FIG. 8-21).

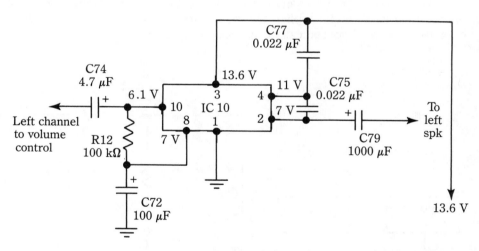

8-21 The separate power-output IC has the audio input at pin 10 and amplified audio at pin 2 of IC10. The left and right channels are identical.

DISTORTED RIGHT CHANNEL

Often, audio output circuits cause most of the audio distortion. Check both the right and left audio signals at the volume controls. If the right channel is distorted, proceed to the right input terminal of the output IC or transistor. Carefully feel the leaky IC or transistor because it will run quite warm. Measure all voltages on the IC or transistor terminals and compare them with the schematic. If the schematic is not available, compare the voltages with those in the good channel.

Check the bias and bypass capacitors that are connected to the IC terminals. A burned resistor or an open or leaky coupling and bypass capacitor can cause audio distortion. Replace all resistors that are burned or have changed values. Remove one end of the capacitor or diodes for an accurate leakage test. Do not overlook a leaky or open output coupling capacitor for distorted, weak, or dead audio conditions.

Test leaky or open push-pull audio output transistors with a DMM. Inspect and test all diodes and bias resistors with the transistor output of the circuit. Make sure that these low values of resistance are replaced with parts that have the exact resistance. Usually, a shorted output transistor has a burned or open bias resistor. Check for leaky diodes in the base circuits of some audio transistor circuits. Audio output transistors and ICs can be replaced with exact universal replacements.

PREAMP AND DOLBY REGULATOR CIRCUITS

The preamp and Dolby IC circuits' supply voltage can come from a transistor regulator circuit. Actually, the 12-V supply that feeds these circuits is isolated away from the B+ power circuits (FIG. 8-22). The power supply or dc battery voltage is fed from the switch through R74 to the collector terminal of Q51.

The 12-Vdc supply voltage is taken from the emitter terminal. When Q51 opens, no voltage can be applied to the preamp or Dolby circuits. If Q51 leaks, a higher-than-normal voltage can be applied to these circuits.

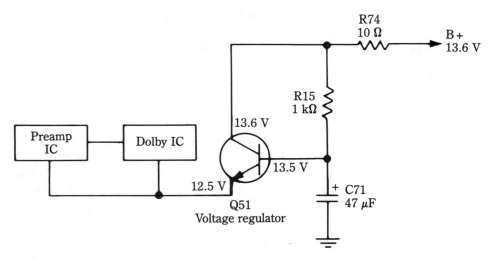

8-22 The preamp and Dolby IC circuits might have a voltage-regulator circuit to supply regulated at 12.5 V.

NOISY VOLUME CONTROL

After several years, the volume control will become noisy as the volume is turned up or down. Sometimes, the audio cuts in and out if the volume control is worn. This can occur in only one stereo channel. Temporarily spray shield or cleaning fluid inside the volume control area. Don't spray fluid into the front end of the control shaft to clean the controls (FIG. 8-23). Remove the car player and the top cover to get at the control. Insert the plastic tip into each volume control section and spray in the cleaning solution. Rapidly rotate the volume control shaft to clean the noisy control. Replace the entire control if it does not work.

8-23   Clean the noisy volume control by spraying shield or cleaning fluid inside the control area of each channel.

HOT IC

The leaky or shorted power IC could run red hot. Spot the discolored IC case and carefully feel the air above the overheated IC. Normal power ICs can run warm, but not red hot.

Often, when an IC shorts, resistors and PC wiring also burns. Sometimes the "A" lead terminal wire with fuse will be burned. The insulation might be charred and this wiring must be replaced. Replace all components that are burned. Repair burned or stripped PC wiring with pieces of hookup wire.

After all burned parts have been replaced, install a new IC. Carefully solder each terminal lead. Now, measure the terminal resistance to the chassis ground. Start with a 10-kΩ range, which can always be lowered. Compare the resistance measurement of each terminal with the normal IC. These comparable resistance measurements should be within a few ohms of each other. Make sure that the DMM resistance numbers stop before marking down the resistance. If the resistance of one measurement is way off, look for additional damaged parts in the hot IC circuits. Then, take accurate voltage measurements on each terminal and compare them (FIG. 8-24).

8-24 Take accurate voltage measurements on the dead channel IC before and after replacing it.

DEAD LEFT CHANNEL

When either channel is dead and the other normal, try to isolate the suspected channel by signal tracing or by tracing the dead speaker wires back to the respective IC. Usually, the stereo IC or transistors are lined to the left and right when looking at the cassette player from the front end. Touch the left channel volume control center terminal with a meter probe with volume wide open. You should hear a loud hum in the speaker.

If you hear no sound, go to the other center terminal and repeat the procedure. When both controls are dead, go directly to the dual output IC and take voltage measurements. The dead left channel can be signal traced with an audio amp or an external signal tracer. Remember, a dead audio channel is much easier to locate than an intermittent one.

Proceed through the audio circuits until the signal is lost. Check the voltage within the transistor or IC terminals. Dead audio circuits are often caused by open or leaky transistors or ICs (FIG. 8-25). Do not overlook the possibility that an electrolytic coupling capacitor might be open—especially capacitors under 10 μF. An open tape-head winding can also produce a dead left channel.

KEEPS REVERSING DIRECTION

In some auto cassette players, when the tape reaches the end of rotation, the procedure is reversed mechanically or with electronic features. The automatic-reverse motion might be activated by a magnet that is attached to the bottom of

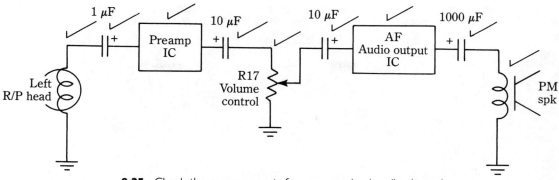

8-25 Check these components for open or dead audio channel.

8-26 The turntable might have a magnet and magnetic switch in the reversing circuits of the cassette player.

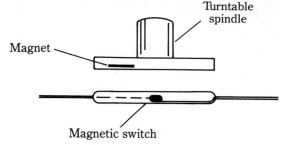

the turntable or reel. Under the turntable is a stationary magnetic switch that operates in the auto-reverse circuits (FIG. 8-26).

The cassette player might quickly reverse directions with a defective switch, broken switch wire, or stationary turntable. When the turntable stops rotating, the switching circuit will automatically keep reversing directions. Check the turntable for a broken drive belt or a missing idler wheel. The turntable must keep rotating in either direction. When the turntable stops, the automatic-sensing circuit energizes a relay, which reverses the direction of the cassette motor.

Some auto cassette players have a commutator ring with spring-like tongs that keep the automatic-reverse circuits operating while rotating. A bent prong, poor prong contact, or dirty commutator ring can result in rapid reverse procedures (FIG. 8-27). Just clean the commutator and tongs with a cleaning stick and alcohol to solve the erratic changing of directions of cassette player.

NO AUTO REVERSE

The auto-reverse procedure can operate mechanically, like the regular cassette player, or electronically. When AMSS is found in the car cassette player, it can operate by detecting the no-signal segment of a tape. Four to six seconds after the recording has ended, the AMSS circuit reverse the direction of the tape (FIG. 8-28).

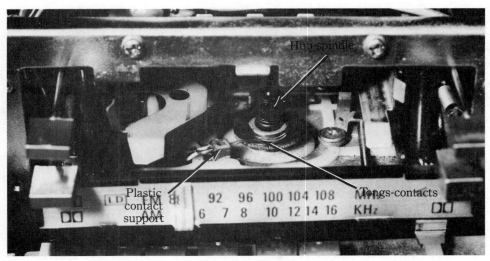

8-27 Clean dirty commutator rings if the cassette player rapidly changes directions.

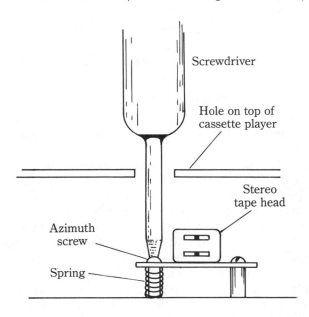

8-28 Take a small screwdriver and adjust the azimuth screw on the tape head while playing a cassette with piano or string music.

Often, dual tape heads, dual capstans, and flywheels are used in cassette players that have auto reverse. The solenoid plunger energizes and pulls in another set of tape heads and the capstan/flywheel assembly, while switching the direction of the dc motor. The AMSS LED indicator and reverse-normal LEDs can be switched in the circuit.

The AMSS circuits can fail to function if the tape heads are dirty or if the tape no-signal segment is shorter than four seconds. The circuits can fail when the tape recording is too low or if a high noise level is found on the tape. Clean the tape heads and adjust the azimuth head before attempting to adjust AMSS circuits. Closely follow the manufacturer's AMSS adjustment procedures.

If the AMSS electronic auto reverse still doesn't function, check all transistors and diodes in the AMSS circuits. Signal trace the audio signal to the input terminal of the IC. Check the solenoid winding for open or broken wires. Manually push the plunger in and out to see if it is stuck or frozen. Check if the solenoid winding cover is burned. Replace complete solenoid if the winding is charred. Accurate voltage and resistance measurements in the AMSS circuits will locate defective ICs, transistors, or diodes.

AUTO HEAD AZIMUTH ADJUSTMENT

The head azimuth can be adjusted mechanically or electronically. Make sure that the tape head is clean. Either clean it with a cleaning stick and alcohol or with a cleaning cassette. Then, insert a cassette with music that contains piano or violin. Adjust the sound from the speakers for the greatest volume at the high frequencies.

The azimuth screw is located on one side of the tape head. When turning the screw, it will align the tape and the head horizontally. The screwdriver can be inserted through a hole in the cabinet (FIG. 8-29). Very slowly turn the screwdriver left or right, until it reproduces the highest frequencies.

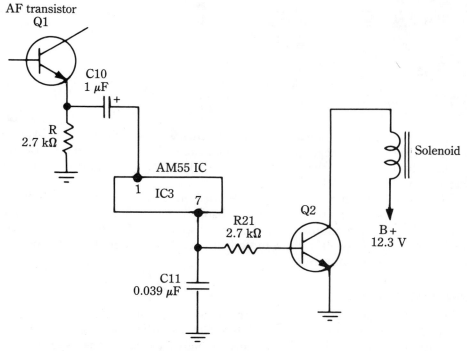

8-29 A block diagram of AMSS audio-detection/auto-reverse circuits.

Electronically adjust the azimuth screw for maximum meter reading with a 1- or 3-kHz test cassette playing. Connect the low-ac meter test probes to one of the channel speaker outputs. Leave the speaker connected and adjust the azimuth screw for maximum loudness and highest meter reading. Only a slight adjustment is needed, unless a new tape head has been installed. If the azimuth adjustment does not make much difference, suspect that either the head is worn, the audio circuit is defective, or that the tape head is magnetized.

MOTOR PROBLEMS

The 12-V cassette motor could operate intermittently, run slow or fast, and appear to be dead. A defective motor can cause excessive noise in the speakers. Monitor the voltage at the motor terminals. Often, the motor is blamed when the motor regulator circuit is actually defective. If the correct voltage is applied at the motor terminals and the motor does not rotate or if it runs slow, suspect that the motor is defective (FIG. 8-30).

8-30 The defective motor might be open, be intermittent, change speeds, or cause noise in the audio circuits.

A defective diode or isolation resistor in series with one motor terminal could be open. Check the voltage on both sides of the resistor and the diode. Do not overlook the possibility that off/on switch that applies voltage to the motor is defective or dirty (FIG. 8-31).

The intermittent motor could operate one hour and not the next. Sometimes, when it is tapped with the screwdriver handle, it will start to run or change speeds. The defective motor can run slow or speed up. Rotating the suspected motor pulley by hand will start the motor running. The dead motor might have an open winding or worn brushes.

LOW SPEAKER HUM

Suspect an open or dry filter capacitor in the dc power supply if you can hear hum when the volume control is turned down. Simply bridge the suspected electrolytic capacitor with another of the same or higher capacity. Make sure that the working voltage is the same or higher (FIG. 8-32).

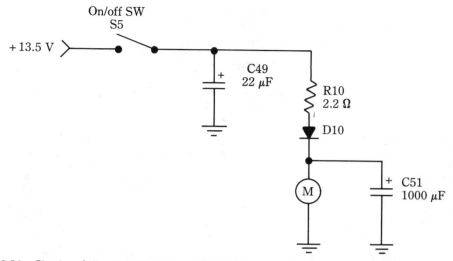

8-31 Check on/off switch S5, R10, and D10 if improper voltage is at the motor terminals.

8-32 Several filter and decoupling electrolytic capacitors are in the power supply of a Toyota 68600-00160 power board.

Lightly tack the filter capacitor in with solder or clip into the circuit with test clips. Never shunt the capacitor while the player is operating or you can damage critical solid-state components. Always check the correct polarity. The positive terminal goes to the positive battery ("A" lead) side.

Shunt all electrolytic capacitors in the power and decoupling circuits (FIG. 8-33). Sometimes real low hum that can be heard by certain people and not by others can be caused with a dry decoupling capacitor. Do not overlook the hum symptom caused by a leaky voltage regulator transistor.

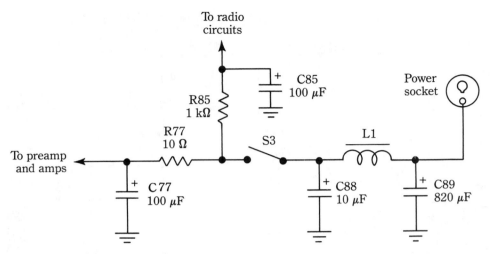

8-33   Shunt the decoupling electrolytic capacitors if a low hum is in the speakers.

Sometimes, when a power output IC has shorted, the PC wiring or choke coil wiring is burned. After shunting all electrolytic capacitors and low hum is still found, suspect a burned or charred choke coil. If the outside paper is scorched or charred, replace the small choke transformer.

Low hum in the audio circuits can be caused by a defective IC or transistor. If the hum is canceled when volume control is turned down, the hum originates before the volume control. Shunt each input terminal of the IC or base of the transistor with a 10-μF electrolytic capacitor to the chassis ground and notice if the hum disappears. Start at the output of the tape head and proceed through preamp and AF stages until the hum is isolated. Hum with weak volume can be caused by a dry electrolytic coupling capacitor.

SPEAKER PROBLEMS

The auto speaker might sound distorted or mushy as a result of extreme weather conditions. The cone begins to warp and the voice coil rides against the center magnet. Excess dirt and dust can fall into the speaker cone and cause distortion or noisy reproduction. Speakers that are mounted upwards in the front dash or rear ledge have a tendency to warp and mush up (FIG. 8-34).

An intermittent speaker might be caused by too much applied power, which damages the voice coil. Sometimes the small flexible wire that is soldered to the voice coil will break and cause an intermittent speaker. Simply remove the speaker from the mounting, connect the speaker wires and lightly press down on the cone of the speaker. Move the cone up and down as the music cuts in and out. Replace the speaker if it is defective. Check the speaker terminals and wires for intermittent audio.

Because at least four speakers are in the new auto hookups, you might have four times more speaker damage (FIG. 8-35). The voice coil could be blown or torn

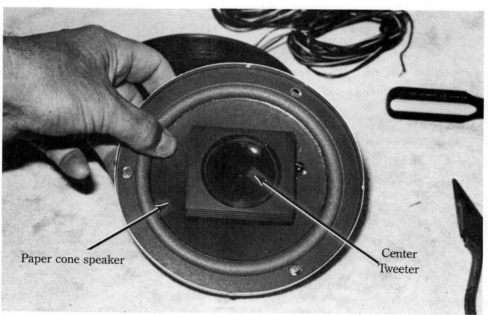

8-34 Press lightly on the speaker cone to locate an intermittent speaker.

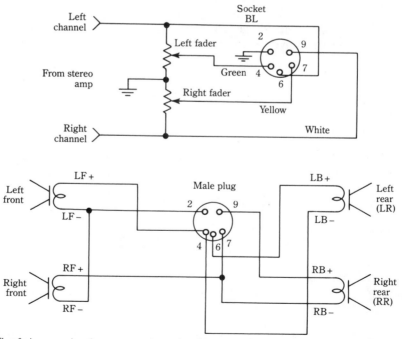

8-35 The fader, speaker harness, and speaker hook up to the front and rear speakers in typical stereo-cassette output circuits.

8-36   Check the voice-coil winding of a dead speaker with the RX1 range of an ohmmeter.

loose from the cone of speaker as a result of excessive volume. Most speakers damaged by excessive volume are large woofers. Check the suspected dead speaker with RX1 ohmmeter range (FIG. 8-36). Replace the defective speaker with universal PM types that have the same wattage, size, and mounting holes.

Chapter 9

Repairing auto cassette/CD players

The auto cassette/CD player might consist of the auto reverse with CD changer controls. The cassette/tuner might have a CD changer in the trunk. The CD changer might operate directly from a cassette/receiver. The under-dash CD system can be connected to the auto cassette player high-power amplifier. Check chapter 8 for auto cassette problems.

The Pioneer KEH-M3000 is one of the lower priced cassette receivers with CD changer controls. You can place the CD changer in the trunk and select any disc or track and program up to 32 tracks for playback. The auto receiver has a quartz Supertuner III with 18 FM and 6 AM presets. The cassette player includes auto-reverse, key-off pinch-roller release with separate bass, treble, and speaker fader controls. The tape frequency response is a +40 to 14,000 Hz with 10 W output per channel.

The Sony XR-7500 cassette/tuner system with dual function controls operates Sony's CDX-A15/CDX-A100 CD changers. The CD changers can be mounted and operated in the auto trunk. This unit can be operated with a wireless remote. The cassette player contains an amorphous tape head, 30- to 20,000-Hz audio frequency, and Dolby B and C to eliminate tape hiss. The preamp cassette/tuner is removable, but it must operate through a separate amplifier.

The Kenwood KRC-930 cassette/receiver contains the finest cassette receiver with the complete operating controls of the 10-disc KDC-C300/KDC-C400 changer. The cassette section has a flat response up to 21,000 Hz, Dolby B and C, full-logic auto-reverse, and programmable tape search. Other features are index, scan, blank skip, and key-off pinch roller release with separate bass and treble controls. Two sets of RCA preamp outputs are provided to an external amplifier. The cassette/tuner can be removed easily and operated with wireless remote.

The Radio Shack 12-1941 under-dash CD player can be connected to a cassette/tuner unit with relay control. The frequency response is from 40 to 20,000 Hz with separate line outputs to an externally connected high-power amplifier or self-contained 36 W per channel with a 4-Ω load. This high-power CD player can be operated independently, plugged into your existing car stereo, or speakers can be added. The player also has automatic play, auto search, two-way audible search, pause control, and LCD display.

SURFACE-MOUNTED COMPONENTS

The latest cassette/CD players have both conventional and surface-mounted components. The surface-mounted components can be mounted on one side of the board with regular parts on the other. Several different boards are found within the cassette/tuner (FIG. 9-1A and 9-1B). Very fine PC wiring connects the surface-mounted processor to the circuit.

Surface-mounted transistors and ICs are used throughout the circuits. Besides surface-mounted ICs and transistors, diodes, capacitors, and resistors are used in many circuits (FIG. 9-2). Notice that diodes can have two active tabs, and the three-legged transistors can have different lead identifications. Power-output ICs should be bolted to heavy heatsinks.

Digital transistors can be found in the audio preamp, and the mechanism and system control. The digital transistor can have an internal resistor in the base circuit or another bias resistor between the base and emitter terminal (FIG. 9-3). When these digital resistors are checked with a transistor tester or the diode test of the DMM, the resistance is higher in the base terminal. Likewise, the measurement from base to emitter with the internal base-to-emitter resistor is different. Compare another similar digital transistor with the low base-to-resistor test before discarding the suspected leaky transistor.

BLOCK DIAGRAM

Before tearing into the CD/cassette player, always check for possible trouble with the block diagram. The block diagram helps to isolate the various stages for easier servicing. The CD mechanism consists of the disc, feed, and load motors with transistor or IC motor drives. Also, the focus and track coils have separate transistor or IC drivers. The servo controller and signal processor with power-control circuits are located in the CD mechanism (FIG. 9-4).

The dc-dc converter supplies +9 V and −9 V with +8-V and +5-V regulators, ICs, or transistors from the 14-V battery source. The +5-V source feeds the motor power on, MOS microprocessor (IC 601), and controller (IC901 and IC902). The +8- and −8-V source is supplied by IC 703 and IC 704. Both +9 and −9-V sources feed the power-, signal-, and system-control circuits. Primarily, the 14-V source is directly applied to line out and power output amplifier ICs.

The D/A converter (IC 606) converts the digital signal to audio and separates the audio into two separate channels. Separate buffer and low-power filter (LPF)

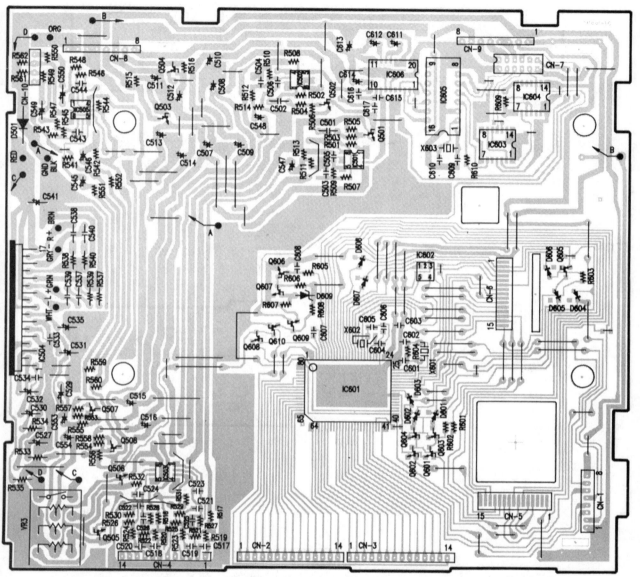

9-1A Surface-mounted components are on one side of the main PC board. Notice the thin PC wiring lines to IC 601. Radio Shack

ICs are found in each stereo channel. The stereo signal is separated after the mute and volume controls and dual-isolator and line amp output jacks. The same stereo audio signal from the volume controls is fed to one dual power amp (IC504). The power amp provides 36 W to each speaker system.

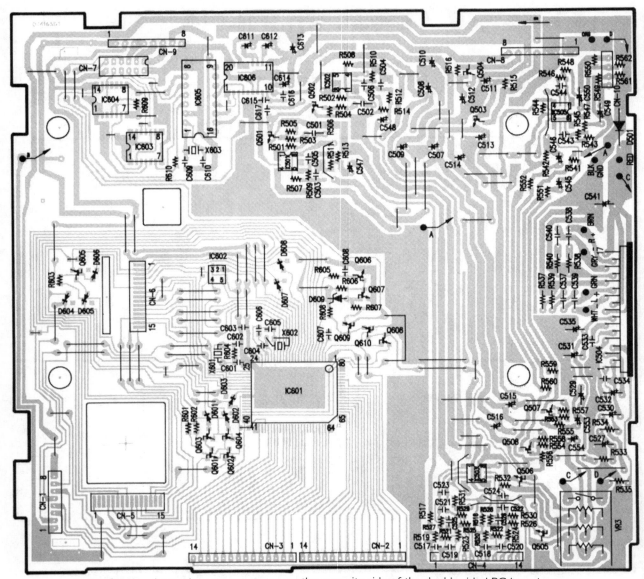

9-1B　Regular electronic components are on the opposite side of the double-sided PC board.　Radio Shack

REMOVING COVERS

The top cover must be removed to get at the various components and boards. In the Radio Shack car compact disc system, remove the six tapping screws that hold the top cover (FIG. 9-5). Then, slip the cover off.

To remove the CD mechanism and the nose piece, remove the clamp wire (B) from the CD mechanism. Pull out the three connectors (A, B, and C) from the

IC AND TRANSISTOR LEAD IDENTIFICATION

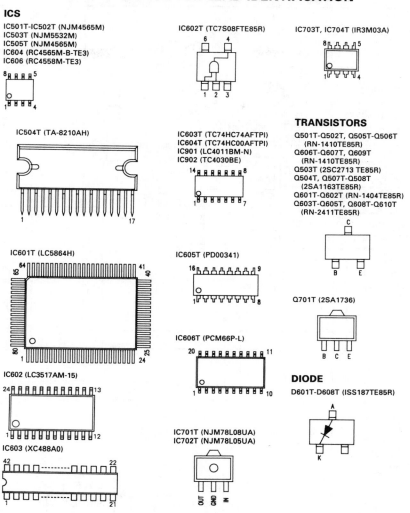

ICS

IC501T-IC502T (NJM4565M)
IC503T (NJM5532M)
IC505T (NJM4565M)
IC604 (RC4565M-B-TE3)
IC606 (RC4558M-TE3)

IC504T (TA-8210AH)

IC601T (LC5864H)

IC602 (LC3517AM-15)

IC603 (XC488A0)

IC602T (TC7S08FTE85R)

IC603T (TC74HC74AFTPI)
IC604T (TC74HC00AFTPI)
IC901 (LC4011BM-N)
IC902 (TC4030BE)

IC605T (PD00341)

IC606T (PCM66P-L)

IC701T (NJM78L08UA)
IC702T (NJM78L05UA)

IC703T, IC704T (IR3M03A)

TRANSISTORS

Q501T-Q502T, Q505T-Q506T
  (RN-1410TE85R)
Q606T-Q607T, Q609T
  (RN-1410TE85R)
Q503T (2SC2713 TE85R)
Q504T, Q507T-Q508T
  (2SA1163TE85R)
Q601T-Q602T (RN-1404TE85R)
Q603T-Q605T, Q608T-Q610T
  (RN-2411TE85R)

Q701T (2SA1736)

DIODE

D601T-D608T (ISS187TE85R)

9-2 The surface-mounted terminals are a little different because they are mounted on a flat surface. Notice that IC504T is a regular dual power-output IC.

9-3 The digital transistors might have an internal resistor in series with the base and bias resistor between the base and emitter.

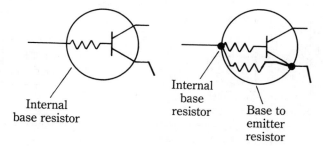

Internal
base resistor

Internal
base
resistor

Base to
emitter
resistor

CD MECHANISM

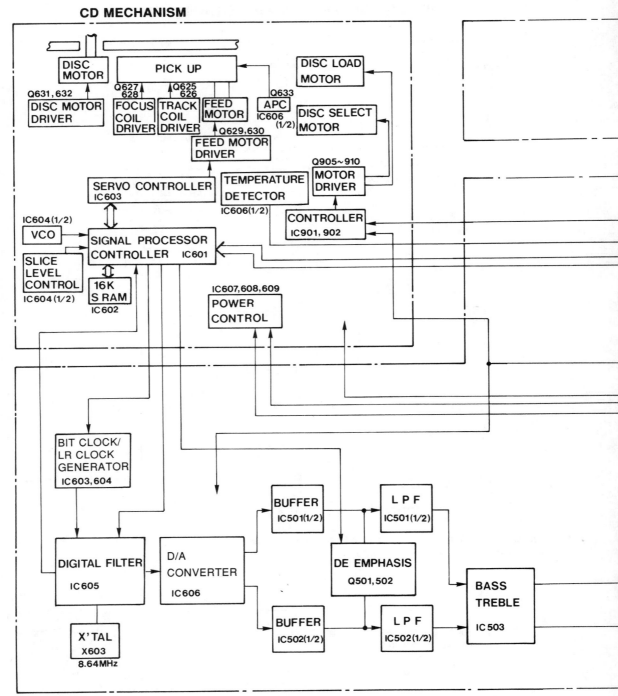

9-4 The block diagram might help to locate the defective part in a certain section of the CD player.

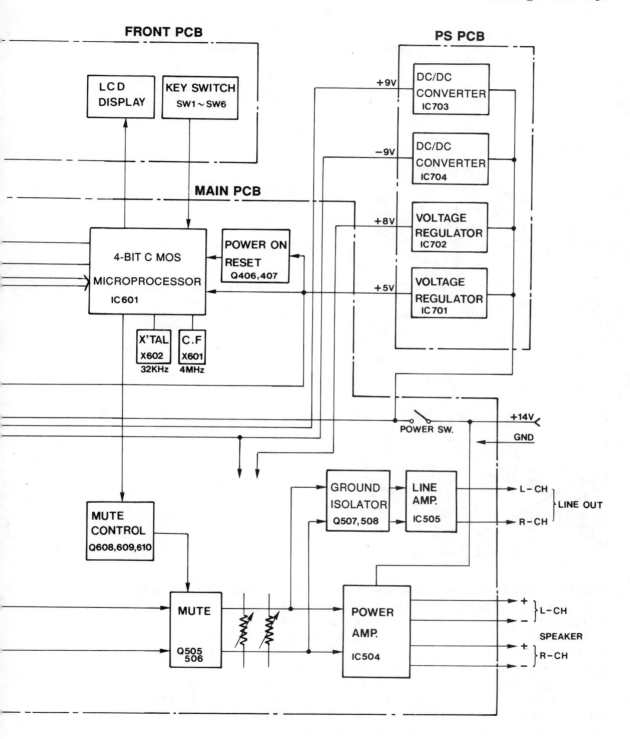

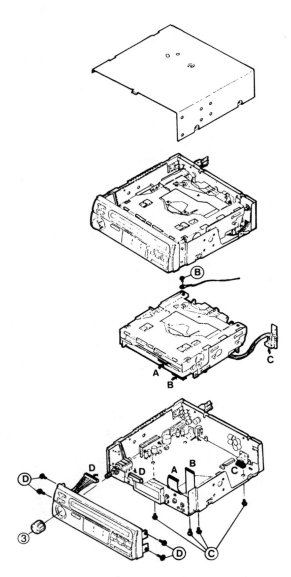

9-5 The covers must be removed to get at the defective section. Here, six metal screws are removed to lift the top cover.

main PC board. Remove the four screws (C) from the bottom cover. Take out the four screws (D) that hold the nose piece. Pull out the connector (D) from the main PC board. Remove the knob (3) from the volume control.

To remove the PC board, take out nut (H) from the volume control (FIG. 9-6). Remove the two screws (E) that hold the power supply PC board. Remove the three screws (F) that hold the front PC board. Take out the screws (G) that hold the cord clamper. Remove screw (I) and screw (J), which hold the heatsink.

Be careful to prevent damage when removing the PC board and the separate components. Place all screws in a saucer or container. Make a list of parts as you remove them. Lay the components in line so that they can be easily replaced after

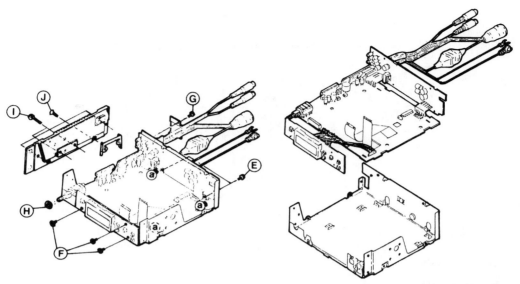

9-6   Remove several components to remove the bottom cover in the under-dash CD player.

you make the repairs. Carefully unsolder the mesh ground straps from the bottom chassis.

SAFETY PRECAUTIONS

When servicing compact disc players or handling static-prone IC processors, give extra care for ICs, processors, and laser-diode pickups, which are sensitive to, and easily affected by, static electricity. If static electricity is nearby, components can be damaged if you do not exercise the proper precautions.

The laser-diode pickup is composed of many optical parts and high-precision components. Extra care must be taken to avoid repair or storage where the temperature or humidity is high, where strong magnetism is present, or where excessive dust is found.

Before attempting to repair or replace any component in the CD player, all equipment, instruments, and tools must be grounded. The CD player should be placed upon a work bench that has a grounded conductive sheet (FIG. 9-7). The metal part of the soldering iron should be grounded. Any repair man or technician should wear a grounded armband.

When working on a CD player, never look directly at the laser beam, and don't let it contact fingers or any other exposed skin. You cannot see this beam with the naked eye. Try to keep a disc on the turntable or keep the laser beam turned away from your eyes.

The laser pickup has strong magnets and should never be brought close to magnetic materials. Keep metal screwdrivers and tools away from the pickup assembly. The pickup should be handled correctly and carefully when removing

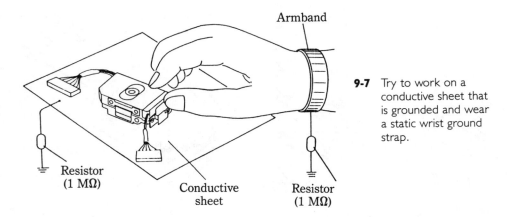

9-7 Try to work on a conductive sheet that is grounded and wear a static wrist ground strap.

or installing. Keep new replacements in a conductive bag until they are ready to be installed. Laser assembly testing and replacement should be done by an electronic technician.

LASER HEAD CLEANING

Erratic operation and complete CD player shutdown can be caused by a dirty laser lens assembly. Wipe the dust off with soft cloth. Dust can be blown off with an air brush, such as those that are used to clean camera lenses. In fact, the camera soft cloth and brush are ideal for cleaning the laser lens assembly. Be very careful not to apply too much pressure because the lens is held by a delicate spring.

TAPE PLAYER AND TUNER OPERATES/NO CD

Check for an open fuse. Inspect the cable harness for poor wiring. Check the "A" battery lead connection. Measure for 14 V at the switched power lead. If it has no voltage, suspect that the power switch is defective (FIG. 9-8). If the CD player is

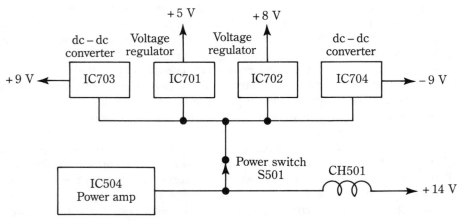

9-8 Check for 14 Vdc across power switch S501. Poor contacts indicate that the switch is defective.

connected to the cassette tuners, does the output relay operate? Listen for the plunger of the output relay. Check for poor choke (CH501) terminals or burned PC wiring.

NO SOUND/FAULTY OUTPUT CIRCUITS

Determine if both channel speakers are dead. If not, clip another PM speaker to the dead channel. When both channels are dead, suspect the dual IC (IC504), the power-supply circuit, or the power switch. Check the dead speaker for an open voice coil or faulty speaker connections. Measure the supply voltage source at the power-output IC.

Often, when the power-output IC becomes shorted, the fuse will blow and the PC wiring might burn. If the CD player has separate stereo line output jacks that have normal sound, the output IC circuits are dead from the volume control to input terminals 7 (left) and 2 (right) of IC504 (FIG. 9-9). Both line output preamp transistors and the input preamp power IC output circuits separate after the volume control.

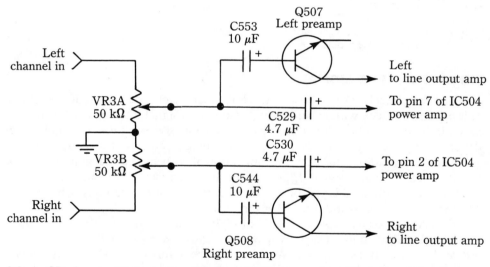

9-9   In CD players with stereo line outputs and high-powered amplifiers, the audio stages are common to the volume controls.

Check the supply voltage at pins 9, 10, and 17 of IC504 if both output channels are dead. Touch one side of C529 and C530 with a screwdriver blade and listen for hum in the speakers. Signal trace the input signal up to pin 2 and 7 of IC504. If the audio signal is normal up to these terminals, suspect that IC504 is defective with normal supply voltage and no sound output (FIG. 9-10). Notice if IC504 is running very warm. Replace IC504 with the original part number or with a universal replacement.

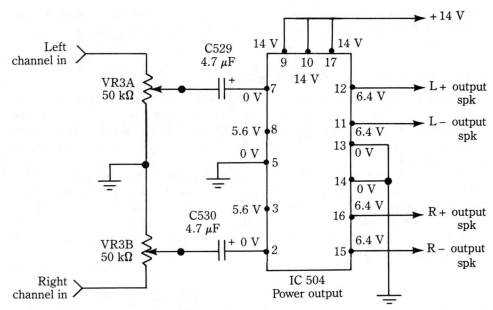

9-10 With a normal audio signal at pins 2 and 7, and not at pins 11 and 12, and 15 and 16, suspect that IC504 is defective.

NO LINE OUTPUT SIGNAL

If the cassette/receiver or CD player has line output jacks, determine if both channels are dead. Check with an audio output amp or a signal tracer at the line output jacks. If the unit has both line-out jacks and separate speaker audio channels, does the speaker circuit operate? Often, only one line power-output IC is defective.

Check the audio signal at the volume control with an external amp or a signal tracer. Trace the audio signal up to preamp line transistor Q507 and IC505 (FIG. 9-11). Usually, a dual IC is the power-line output for both channels. Check for a positive 6.5 V at pin 8 and −6.5 V at pin 4 of IC505.

Check the positive voltage regulator (Q503) and the negative voltage regulator (Q504) for the emitter output voltage. Both voltage regulators are fed from the +8- and −8-V sources (FIG. 9-12). An open voltage-regulator transistor can have no voltage output, but the leaky regulator might have an increase in voltage.

UNGROUNDED SPEAKER OUTPUTS

When installing or repairing the CD-player speaker connections, notice if any of the speakers are grounded. With regular auto radios and cassette players, one side of the speakers are grounded. Here, both channels are above ground and work directly out of the dual audio output IC (FIG. 9-13). If you ground one side of the speakers you can damage the speaker and IC504. Remember, a dc voltage is at these speaker terminals.

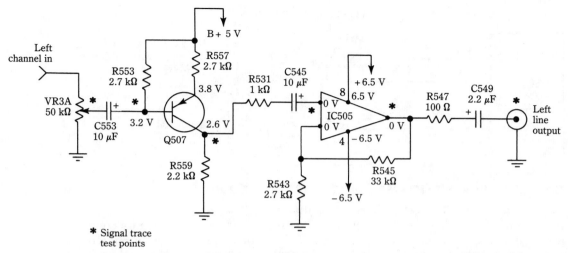

9-11 Signal trace the audio from the volume control (VR3A) to the line-output pins with the audio external amp or signal tracer.

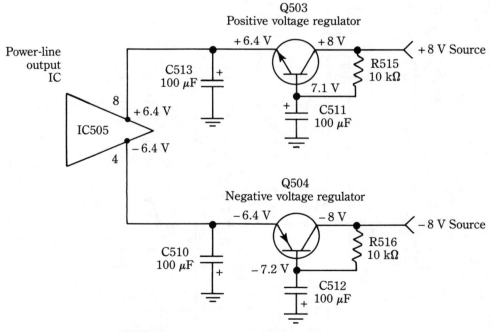

9-12 The positive and negative voltage (6.4 V) from the voltage regulators is applied to pins 4 and 8 of line output IC505.

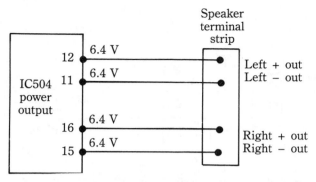

9-13 Notice in this speaker power-output connection that one side of the speaker is not grounded. IC504 damage might occur if the speaker is accidentally grounded.

CASSETTE PLAYER NORMAL/INTERMITTENT CD AUDIO

In some units, the CD player can have an output relay that switches the radio/tape player into the speaker system when the CD player is off. The relay is only energized when the power switch is turned on. The cassette player speaker output terminals are switched to the speakers when the CD player is turned off. Four sets of relay points route the cassette speaker outputs to the speakers (FIG. 9-14). Both the cassette and the CD player have their own internal power amplifiers.

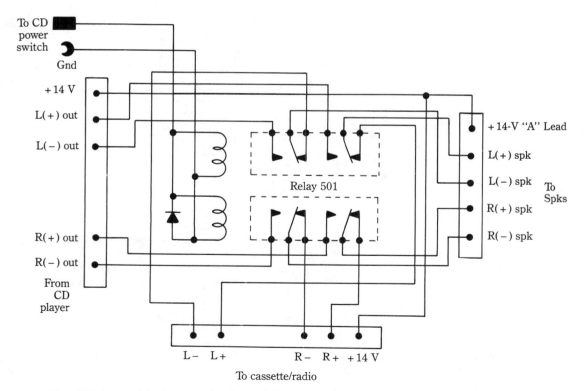

9-14 The CD player might have a relay that switches the speaker connecting from cassette/receiver and CD player to the connecting speakers.

If the cassette player speakers are normal and you find intermittent sound when the CD player is switched on, suspect that the relay contacts are dirty. Notice if the relay solenoid is energized. The 14-V source that feeds the relay comes from a detachable cable in the CD player. Remove the cover and clean each solenoid contact with a piece of cardboard. Sandpaper can help, but clean out all points after cleaning them. Remember, the points that make contact in CD player mode are high until the solenoid is energized. Do not overlook the possibility that a bad cable wire might be leaking to a dead speaker.

WEAK SOUND IN LEFT CHANNEL

Signal trace the left channel at the volume control and compare it with right channel. If both channels are the same, proceed toward the speaker output terminals. Check the audio in both channels at the line output terminals. When both line signals are normal, suspect a defective power output circuit. Actually, the audio signal can be traced from the output D/A converter toward the speakers (FIG. 9-15).

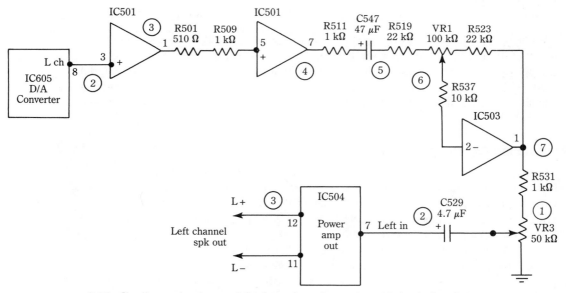

9-15 Signal trace by the numbers from the volume control in both directions.

Signal trace the audio from pin 8 of the D/A (IC605) toward the volume control with an external amp or an audio signal tracer. Check the input and output signal of each IC. Signal trace each side of an electrolytic coupling capacitor. Divide the audio circuit in half at the volume control. If the left channel is weak at the volume control, start at pin 8 of IC605 and proceed toward the volume control. When the signal is normal with both channels at the volume control, proceed toward the speaker output circuits.

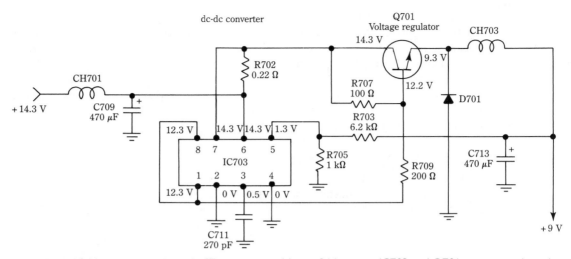

9-16   The 14.3-V source contains a dc CD converter with a +9-V output. IC703 and Q701 connect and regulate the 9-V source.

Check the circuits where the signal becomes weak. Weak audio can occur as a result of bad coupling capacitors, ICs, or transistors. If the signal stops at one of the IC preamps, check the input and output terminals. Likewise, when signal stops at a transistor circuit, test the suspected transistor. Do not overlook muting transistors and ICs. A leaky muting transistor can partially close the audio signal to ground.

DEAD DC-DC CONVERTER

Usually, dc-dc converter circuits convert a high dc voltage down to a low voltage. Here a positive and negative IC dc-dc converter provides a +9 V and −9 V to the various circuits. When one or both of these circuits become defective, the supply-voltage source cripples the connected circuits. Always check the different voltage sources to locate the defective circuit.

IC703 converts a +14-V source to a +9-V source, and IC704 supplies a −9-V source. Both of these ICs are connected to the +14-V source. The +14-V battery supply is wired to terminal 6 with Q701 as a voltage output regulator (FIG. 9-16). When the +9-V source is low or missing, suspect IC703 or Q701. If Q701 opens, no voltage is found at the +9-V source. If it is leaky with a short between the emitter and collector, the output voltage can be higher. However, an internal leak between the emitter and base will lower the output voltage to only a few volts. Remove Q701 if you suspect that it is leaking. Take critical voltage measurements on IC703 if the +9-V source is low or improper.

The negative 9-V source contains only IC704 as the dc-dc converter. The positive 14 V from the battery feeds to pin 6 of IC704 and the negative 9 V is supplied from pins 2 and 4 of ground potential (FIG. 9-17). The leaky IC704 can lower the output voltage; an open IC will have very little negative voltage. Notice that D702

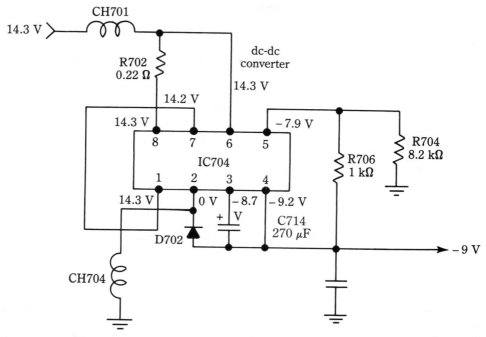

9-17 The negative 9-V source is converted with dc/dc IC704 with −9-V source at pin 4.

and C714 connections are reversed to common ground. Both IC704 and IC703 are identical ICs, but convert opposite voltages with the connection to ground.

NO +5-V SOURCE

Check the +5-V source if you find distorted sound or no sound in the CD output circuits. The 5-V source feeds D/A converter IC607 and also the system-control circuits. IC702 provides the +5-V source (FIG. 9-18). Pin 3 of IC702 receives +14 V from the battery source and the +5-V output at pin 1. Pin 2 is grounded. A leaky or open IC702 can cause no sound or low distortion. Take voltage measurements at all three pins. Replace IC702 if its output voltage is low or improper.

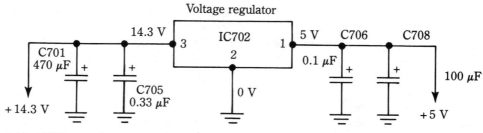

9-18 IC702 provides a +5-V regulated source from the 14-V battery.

NO +8-V SOURCE

Voltage regulator IC701 converts +14 V to a regulated +8 V, which feeds the line output amps, preamps, and mute circuits. The +14-V battery voltage is connected to pin 3 and +8-V output at pin 1 of IC 701 (FIG. 9-19). A leaky or open IC701 can produce a low-voltage (+8 V) source.

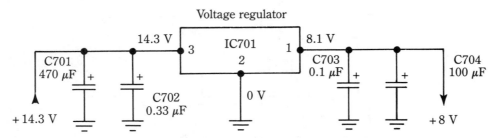

Voltage regulator

9-19 The +8-V regulator (IC701) converts the +14-V battery source to 8 V.

ERRATIC LOADING

Suspect that either improper voltage is at the loading motor, or that the motor or motor control (IC901) is defective with erratic or no loading. The loading motor driver can be a transistor or an IC. Q908, Q909, and Q910 provide a positive voltage to the loading motor. Q909 and Q910 apply a negative voltage to the motor circuits. Q903 and Q904 are voltage regulators for the loading motor circuits (FIG. 9-20). No voltage is found at the motor until the motor control is turned on.

   Check Q903 and Q904 for correct voltage applied to the loading motor circuit. Leaky or open transistors could apply a greater voltage or no voltage to the loading motor drivers. An open Q909 or Q908 could prevent voltage from reaching the loading motor. If the voltage is present at the loading motor terminals and it isn't rotating, suspect that the motor is open. Check motor continuity with RX1 range of the DMM. Inspect the drive belts or gears for erratic speeds. Check if the loading mechanism is defective.

DISC-SELECT MOTOR PROBLEMS

The disc-select motor positive and negative regulated-voltage source feeds from Q901 (+) and Q902 (−). Q905 provides positive disc-select motor-drive voltage and A907 and Q906 apply the negative voltage to select the motor at the pin 11 terminal of C5903 (FIG. 9-21). When the motor will not go forward, suspect Q905 and Q901. For reverse motor operation, check Q906, Q907, and Q902. The negative and positive voltage that is applied to the disc-select motor is controlled by motor control IC901.

NO SPINDLE OR DISC ROTATION

The spindle motor can be controlled by transistors or ICs. Here, the disc-motor voltage is fed from Q632 and Q631. Surface-mounted disc control IC603 provides

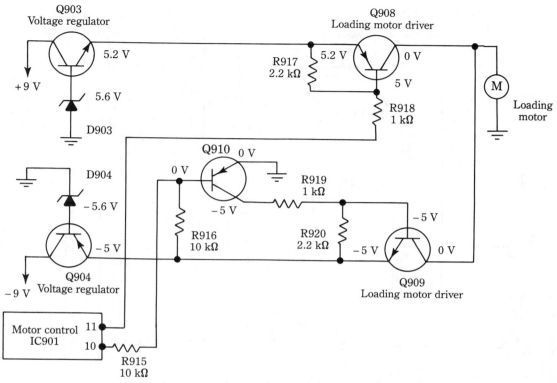

9-20 Motor control IC901 controls the loading motor from a positive and negative 9-V source with loading-motor transistor drivers Q908, Q909, and Q910.

voltage to driver transistors Q632 and Q631. All three motors are fed from common PC board CS903 (FIG. 9-22).

Check the dc voltage at the disc-motor terminals. If voltage is present without rotation, check the motor continuity. Replace the motor if it is open. If an improper voltage or if you find no voltage at the disc or spindle motor, measure the voltage output at Q632 and Q631 (FIG. 9-23). Make sure that a +9 and −9 V is found at the collector terminals. If an improper voltage is at the collectors, check the switching-power voltage at IC607 and IC609. Do not overlook the possibility that the disc mechanism might be defective.

NO FEED-MOTOR ROTATION

The feed motor moves the laser assembly perpendicular to the disc assembly. The dc-voltage feeding motor comes from feed-motor driver transistors Q629 and Q630. The feed-motor drivers can be transistors or ICs. Check for both −9 and +9 voltages at the collector terminals of the driver transistors. Suspect that Q629 or Q630 is defective if no voltage is at the feed-motor terminals. Measure the continuity of the feed-motor terminals for an open motor with the RX1 range of the DMM.

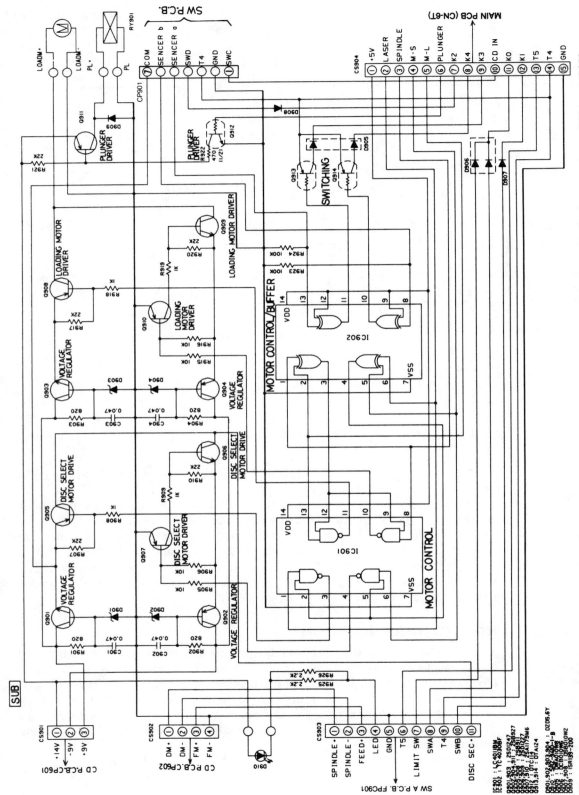

9-21 Motor control IC901 controls disc-select transistors Q906 and Q907 with a positive and negative voltage from voltage regulators Q901 and Q902.

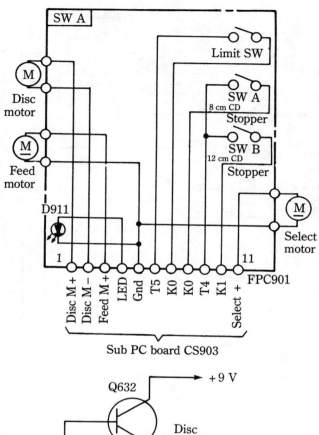

9-22 The disc, feed, and select motor terminals are taken from sub PC board CS903. The voltage fed to the motors can be checked at this PC board.

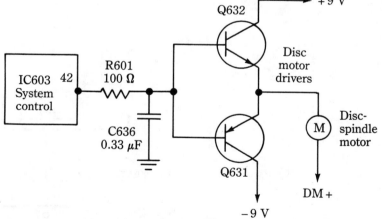

9-23 Driver transistors Q631 and Q632 provide voltage to the spindle or disc motor.

WIRING DIAGRAM

It is very difficult to troubleshoot the various CD player circuits without a circuit or wiring diagram. The various PC boards and units are wired together with a different wiring diagram (FIG. 9-24). The speaker cables, power leads, and line output plugs are connected to the main PC board.

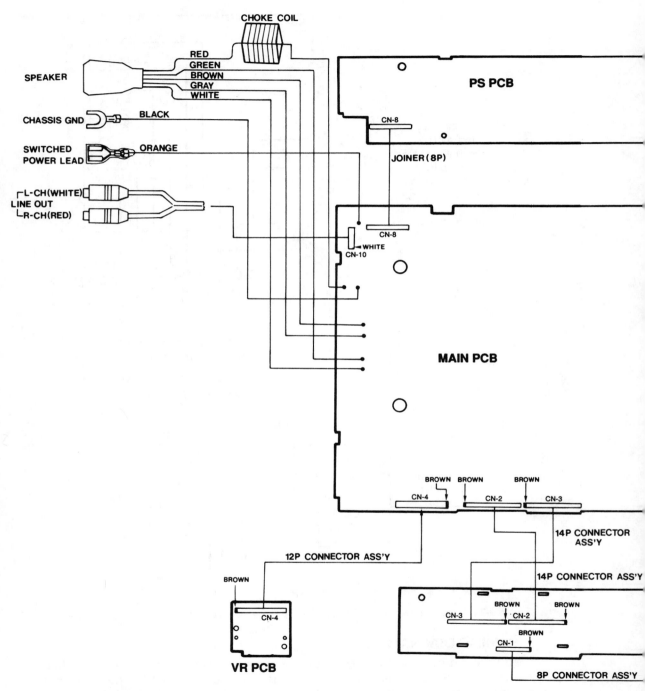

9-24 The wiring diagram shows how the speaker cable, line-out jacks, and wiring connects the various boards together. Radio Shack

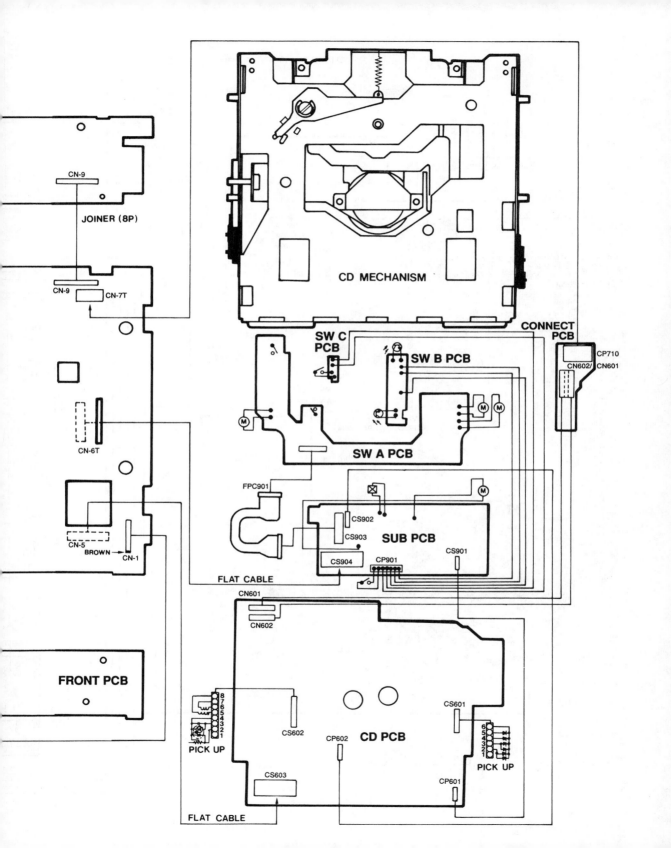

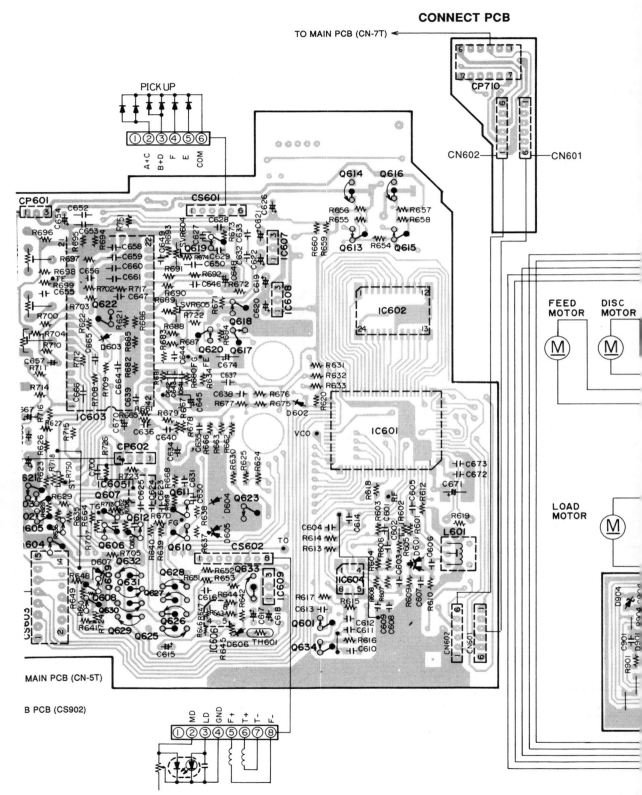

9-25 This wiring diagram shows how the different motors and sensors are connected in the circuits.

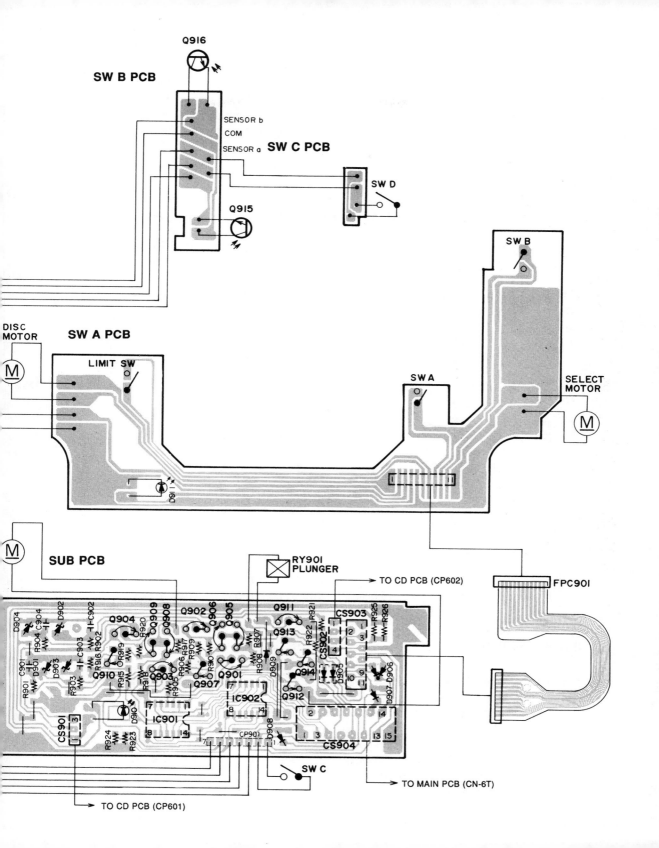

The feed-, disc-, and select-motor connections are on the SWA PC board with the disc-detect and disc-in sensors (Q916 and Q915) on the SWB PC board (FIG. 9-25). The loading motor is fed from the sub-PC board. If a schematic wiring diagram is not available, all leads must be traced to the respective boards, which wastes a lot of time.

IMPROPER SEARCH

With search problems, check the tracking-servo, kick-pulse, and feed-motor circuits. A defective tracking-servo circuit could be caused by Q625 and Q626. Measure the voltage on both transistors. Check the TP and TE waveforms. Test both transistors for opens or shorts (FIG. 9-26). Measure the voltage on tracking IC605. The tracking driver can have transistors or ICs in the CD chassis. Check the waveform at pins 8 through 11 of IC603 and at pins 22 and 33 of IC601. Check all voltages in the kick-pulse circuits. Measure voltage at pins 0 through 8 of the servo processor (IC603) and at pins 22 and 33 of the signal-processor control (IC601).

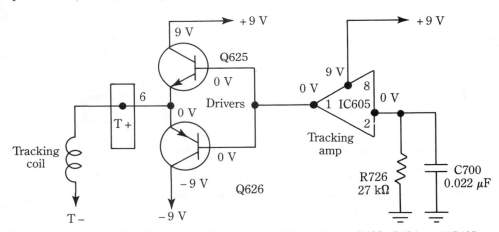

9-26   The tracking coil is driven with voltage and signal from drivers Q625, Q626, and IC605.

Check the feed-motor circuits. Measure voltage in Q629 and Q630 (see FIG. 9-27). Test Q629 and Q630 for open or shorted conditions; remove them from circuit, if in doubt. Check R715 for opens or burns.

EXTERNAL WIRING

The cassette/tuner, cassette/receiver, or underdash CD player is connected together. The cassette tuner or cassette receiver with CD controls operates on a CD changer within the auto trunk. The line output of the CD player connects to the line out of the cassette receiver/tuner. Both the cassette/tuner and the CD player "A" leads must be connected to the 14-V battery source. If the CD player has internal stereo power output stages, they are connected to the speakers (FIG. 9-28). When neither unit has power-output amps, the cassette/receiver must be connected to the external power amplifier and speakers.

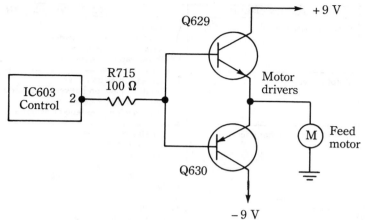

9-27 Check the voltage at the disc motor supplied by transistors Q629 and Q630 from a 9-V source.

A CD under-dash troubleshooting chart is shown in FIG. 9-29, and a schematic diagram must be used to locate the various parts within the troubleshooting chart. Figures 9-30 and 9-31 show the CD schematic and main diagram.

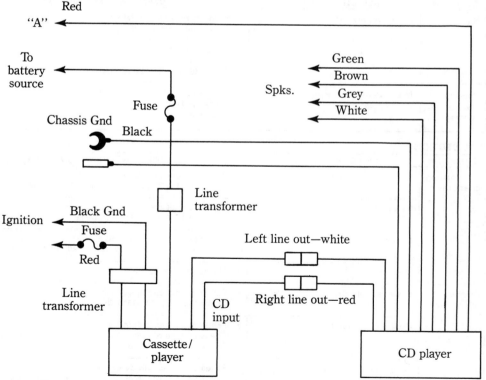

9-28 The external wiring connections of a cassette/receiver and under-dash auto CD player.

SYMPTOM	DEFECTIVE CIRCUIT	DEFECTIVE POINT AND CAUSE
No sound	Power supply circuit	• Fuse open. • Faulty connection between battery. • Power switch defective. • Check lead wire cold soldered.
	Output circuit	• Speaker voice coil open. Faulty connection between speaker and connection core. • Power amplifier defective. Check each pin voltage of power amplifier IC504 and IC505.
	Control flat amplifier circuit	• Variable resistor VR1, VR2, VR3. • Line amplifier defective. Check voltage of IC505. • Voltage regulator defective. Check voltage of IC701, IC703, Q701, Q503, Q504. • BASS TREBLE circuit defective.
Distorted sound or insufficient sound	Output circuit	• Speaker wire grounded. • Power amplifier defective. • Check each pin voltage of power amplifier IC504, C529-C532, C535, C537-C540, R537-R540 defective. • BASS TREBLE circuit defective. • Line amplifier defective. Check voltage of IC505.

DISPLAY SECTION

SYMPTOM	DEFECTIVE CIRCUIT	DEFECTIVE POINT AND CAUSE
No display	Power supply circuit	• Power supply circuit defective. Check voltage of IC702.
	Micro computer circuit	• Micro computer circuit defective. Check IC601. Check X601, X602, Q606, Q607, Q609.
Missing display segment	Micro computer circuit	• Micro computer circuit defective. Check IC601. • LCD defective.

DISC SECTION

SYMPTOM	DEFECTIVE CIRCUIT	DEFECTIVE POINT AND CAUSE
Disc loading inferiority	Power supply circuit	• Power supply circuit defective. Check Q901-Q904, D901-D904 and R901-R904 short or open.
	Sensor circuit	• Sensor circuit defective. Check Q915, Q916, D910 and D911.
	Motor driver circuit	• Motor driver circuit defective. Check IC901, IC902, Q905-Q910, R905-R910 and R915-R920.
	Mechanism	• Mechanism defective.
Disc turning inferiority	Power supply circuit	• Power supply circuit defective. Check IC607-IC609, Q613-Q616.
	Feed motor circuit	• Feed motor circuit defective. Check voltage of IC603, 2 pin and IC602, 25, 26 pin. Check Q629, Q630 short or open.
	Focus search circuit	• Focus search circuit defective. Check voltage of IC603, 35 pin and IC601, 13 pin. Check voltage of IC605, 7pin. Check Q627, Q628 short or open.

9-29   The CD-player troubleshooting chart, used in conjunction with a schematic diagram.

SYMPTOM	DEFECTIVE CIRCUIT	DEFECTIVE POINT AND CAUSE
Disc turning inferiority	Automatic power control circuit	• Automatic power control circuit defective. Check Q633 short or open. Check voltage of IC606, 1 pin.
	Disc motor circuit	• Disc motor circuit defective. Check voltage of IC601, 11pin and IC603, 40, 42 pin. Check Q631, Q632 short or open.
	VCO circuit	• VCO circuit defective. Check frequency of TP VCO and readjust L601. Check voltage of IC604, 1-3 pin. Check waveform of IC601, 2-3 pin.
	Focus servo circuit	Check voltage of TP FE and readjust SVR605.
Search inferiority	Tracking servo circuit	• Tracking servo circuit defective. Check waveform TP, TE, and readjust SVR601, 602, 603. Check Q625, Q626 short or open. Check waveform of IC603, 9-11 pin and IC601, 19-21 pin.
	Kick pulse circuit	• Kick pulse circuit defective. Check voltage of IC603, 0-8 pin and IC601, 22, 33pin.
	Feed motor circuit	• Feed motor circuit defective. Check Q629, Q630, R713 short or open.
No sound	Digital signal circuit	• Digital signal circuit defective. Check waveform of IC602, 1-24 pin. Check waveform of CN601 defective IC601.
Noise	VCO circuit	• VCO circuit defective. Check frequency of TP VCO and readjust L601. Check voltage of IC604, 1-3 pin. Check waveform of IC601, 2-3 pin.
	Slice level control circuit	• Slice level control circuit defective. Check voltage of IC604, 5-7 pin.
	RF circuit	• RF circuit defective. Readjust SVR604.
	Mechanism	• Mechanism defective. Check eccentricity of mechanism.
Disk eject inferiority	Sensor circuit	• Sensor circuit defective. Check Q915, Q916, D910 and D911.
	Motor driver circuit	• Motor driver circuit defective. Check IC901, IC902, Q905-Q910, R905-R910 and R915-R920.
	Power supply circuit	• Power supply circuit defective. Check Q901-Q904, D901-D904 and R901-R904 short or open.
	Mechanism	• Mechanism defective.
No sound	Mechanism	• Mechanism defective.
	RF circuit	• Digital filter, D/A converter, LPF circuit defective. Check IC605, IC606, IC501 and IC502.
	Display circuit (Micro-computer)	• Micro-computer circuit defective. Check IC601.
Distorted sound or insufficient sound	RF circuit	• Digital filter, D/A converter, LPF circuit defective. Check IC603, IC604, IC605, IC606, IC501 and IC502.
	Mechanism	• Mechanism defective.

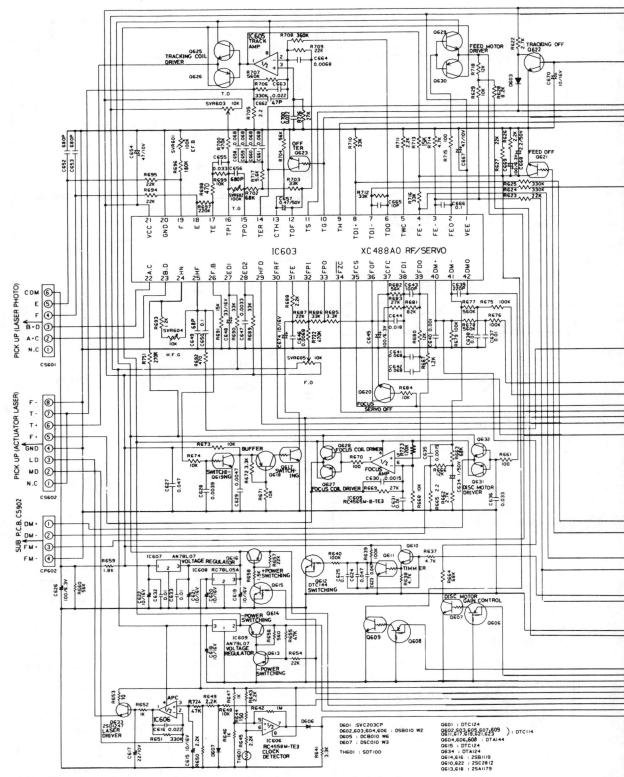

9-30 The schematic diagram of the servo and signal processor-control circuits.

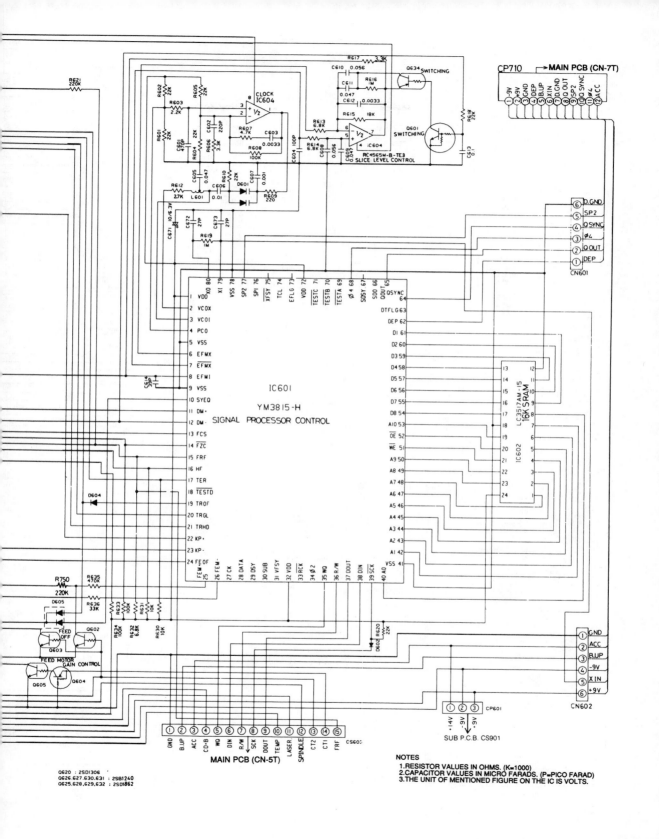

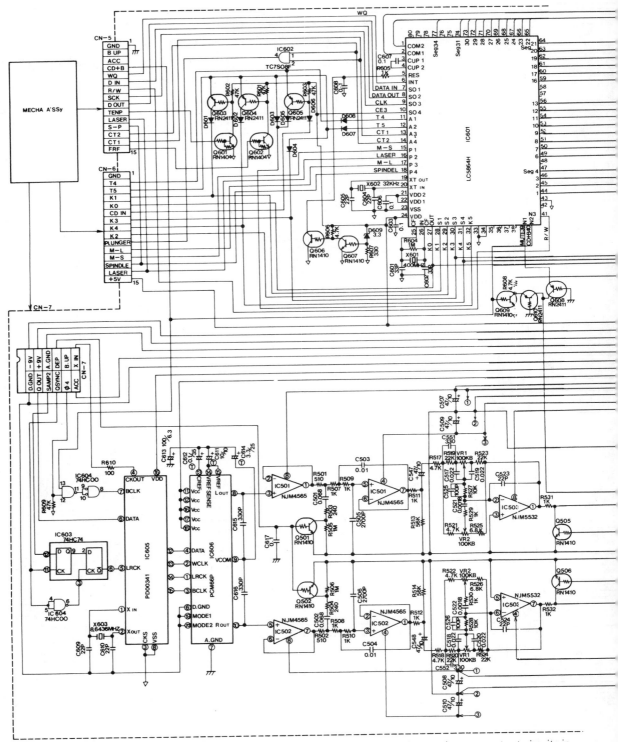

9-31 The main schematic diagram of the RF signal, D/A converter, and audio line- and power-output circuits in the under-dash CD player.

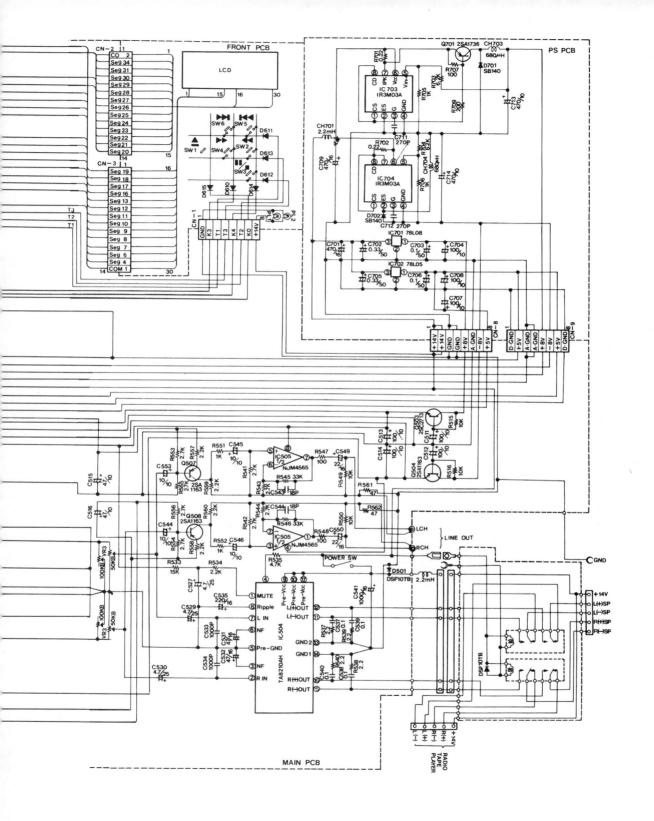

Chapter **10**

Maintaining double cassette decks

*T*he dual-deck cassette player is found in boom boxes, cassette decks and cassette CD players. The decks might be referred to as deck A (or number 1) and deck B (or number 2). Some AM/FM/MPX cassette portables offer a dual cassette deck with auto-reverse playback on deck A, and record/play on deck B. Other dual decks have auto-reverse and record on deck A and playback on deck B. Some dual cassette decks offer relay play and high-speed dubbing (FIG. 10-1).

The deluxe dual cassette deck might have up to 10 W of power. In 2-way/4-speaker system, 4 detachable speakers are found in larger cassette portables. Large 6-inch woofers and 4-inch subwoofers are used in the deluxe models. Most dual cassette decks can be operated from ac power or batteries.

NO ROTATION

Rotate the function switch to all other operations when the cassette deck will not rotate. Check both decks for tape rotation. If nothing is working, go directly to the power supply or batteries. Does the cassette operate on ac and not batteries or vice versa? Check the on/off switch when nothing functions. Inspect the schematic for an open fuse. Check and replace any dead batteries.

Check the output voltage at the on/off switch (FIG. 10-2), The on/off switch (S6) is located after the ac/dc switch and the external dc output jack. Clean any dirty switch contacts. Do not overlook the possibility that the contacts of a remote control jack, which are in series with the on/off switch in some remote units, might be defective.

10-1 The double cassette deck might be located in a portable that has removable speakers.

When the cassette deck operates from batteries and not ac, suspect that an ac circuit is defective. Measure the dc voltage across the large filter capacitor. If voltage is found here and not at the on/off switch, suspect that the ac/dc switch is defective or dirty. If you find no voltage at either switch, the ac rectifier circuit is defective.

Check each diode in the full-wave or bridge rectifier circuit with the DMM diode test. A shorted or leaky diode might damage the power transformer (T1). Check the primary winding with the ohmmeter (FIG. 10-3). No resistance measured indicates that the primary winding is open.

Sometimes, these small ac-powered transformers can be repaired by removing them from the circuit. You might find that the power transformer is mounted off to one side of the PC board (FIG. 10-4). Carefully cut one side of the transformer's heavy wrapping next to the lamination. Select the side where the ac wires enter the transformer winding. Bend and pull the wrapping back to get at where the wires are soldered.

Carefully pull each ac wire upward. Sometimes the small coil wire is burned off where it was soldered together at the ac junction. You might find that one of the coil wires was not soldered or that the joint was soldered poorly. Scrape off the enameled insulation at the coil wire end and tin it with solder paste. Reconnect the coil wire to the ac wire and solder them. Apply solder paste to this joint for a good job. Now, check the continuity of the primary winding with RX1 ohmmeter scale. Replace the transformer if the wires are broken inside winding area.

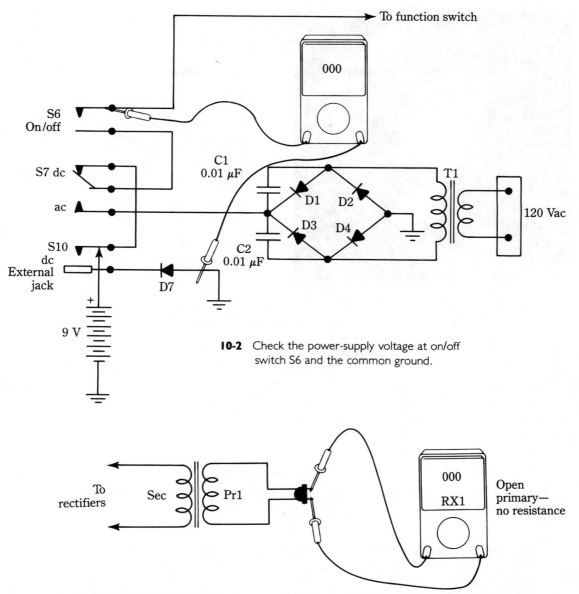

10-2 Check the power-supply voltage at on/off switch S6 and the common ground.

10-3 Check the continuity of the primary winding of T1 if no ac voltage is at the bridge rectifiers.

KEEPS BLOWING FUSES

In the larger portables, you might find a fuse to protect the power supply circuits. Check the silicon diodes when the fuse keeps blowing. If any diode shows leakage inside the bridge rectifier, replace entire unit. The bridge rectifier can be replaced with four single diodes (FIG. 10-5).

10-4   The small power transformer might be off to the side and mounted in the plastic cabinet.

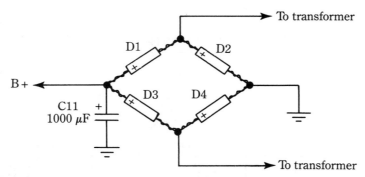

10-5   If the correct bridge rectifier is not available, select the correct single diodes and wire them together.

Simply connect each diode in series with the two negative (anode) ends to ground and the positive (cathode) connections at B+. The ac secondary transformer winding is connected where the last two diodes connect. If one diode is reversed, the diode might open or run warm. Check each diode for correct polarity with the diode test of the DMM. Here, the positive (red) terminal of the meter connects to the anode terminal of the diode and the negative meter probe connects to the collector of the diode for a normal measurement. The reverse is true when using a regular VOM (FIG. 10-6).

Sometimes the fuse will hold after replacement. Either one of the diodes might have flashed over or a power-output IC might have arced-over momentarily.

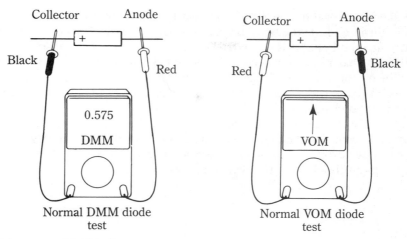

10-6 Make a normal DMM diode measurement by placing the black probe to the collector and the red probe to the anode (−) terminal.

In this case, the fuse might hold indefinitely. Always check across the filter capacitor for leakage, after replacing the diode. Recheck the B+ circuits if you find a very low ohmmeter reading.

NO DECK 2 ROTATION

When one of the dual-cassette decks will not rotate, check both decks. Both decks might be dead if one motor operates both decks from one dc motor (FIG. 10-7). Often, each dual tape deck has its own motor.

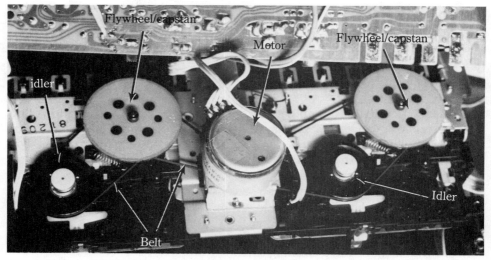

10-7 In this GE dual-cassette deck, one motor drives both capstan/flywheels.

Make sure that the motor drive belt isn't loose or off the motor pulley. If the motor is turning inside the belt, the belt is too large or the flywheel/capstan is frozen. Replace the belt if it is defective.

Suspect that a motor is dead if the drive belts are okay. Check the dc voltage across the motor terminals. Replace the motor if you find voltage across the terminals. Try to rotate the motor by hand. If it takes off, replace it anyway. In time it will become intermittent.

Check the capstan/flywheel for dry bearings. Remove the flywheel if the unit is frozen and cannot be rotated. Clean the capstan bearing hole with alcohol and a stick. Just put a drop of light oil or grease on the capstan bearing. Readjust the end play of capstan/flywheel in the deluxe models (FIG. 10-8).

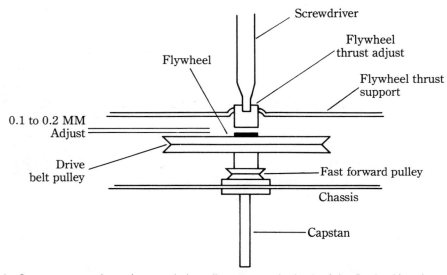

10-8 Some cassette players have end-play adjustment at the back of the flywheel bearings.

EATS TAPES

When a cassette deck eats or pulls tape from the cassette, check the rotation of the take-up reel. A slow or erratic take-up turntable will let the tape spill out and wrap around the pinch roller and capstan. Remove the cassette and the excess tape. Sometimes the tape can be threaded back into the cassette with a pencil in the take-up turntable sprocket.

Try another cassette. Sometimes a sticky substance on the tape might cause it to spill out or get eaten. Excess oxide on the capstan and pinch-roller assembly might pull the tape. Clean all rotating surfaces and try another cassette. If the take-up spindle is erratic, check the driver idler or belt for slippage. Clean all rotating surfaces with alcohol and a cleaning stick.

BROKEN PLASTIC DOOR

In many of the dual-cassette decks, the front plastic door will come off. Lightly pry up the front door. The front door slips into a four-sided track that clips the plastic door to the front-loading assembly (FIG. 10-9). Sometimes, one or two of these plastic tips become broken and the door falls off. Apply cement to the broken plastic ends and slip the door back into position.

10-9 Most plastic doors slip off of the front-door assembly. Repair them with model glue.

Hinged doors on some models will unclip at the bottom. Lightly pull up on the door assembly and see if the door will loosen. Hinged doors that have long treaded rods must be replaced when one or both plastic ends are broken. A metal section of the front plastic door might be removed with two etcheon-type metal screws. The plastic door should be removed to easily clean heads and remove jammed cassettes.

BROKEN PUSHBUTTONS

On vertically mounted tape decks, the pushbutton end that fits into the metal lever might be pushed too hard and break if the knob will not move. Sometimes the unit is dropped or side-swiped and a knob is broken. Often, the knob lever at the rear is broken (FIG. 10-10). This type of knob must be replaced and cannot be repaired.

After ordering and securing the new knob, remove long rod that holds all knobs in position. Usually, a ''C'' washer holds the rod into place. Remove the ''C'' washer that is close to the broken button assembly. The cassette assembly must be removed to get at the front button assembly. Pull out the rod and lay each

10-10   Replace the entire button assembly if the back tip is broken off of the button.

button into a line for replacement. Only remove the part of the rod up to the broken button. Replace the broken button, and place the others in a straight line (FIG. 10-11). Replace the "C" washer and make sure each knob operates properly.

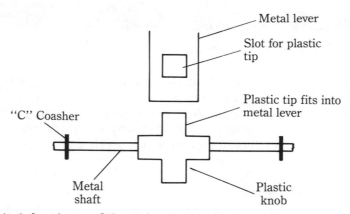

10-11   The plastic front buttons fit into a slot; if the metal lever is jammed, the tip will break off.

PLAYS ONLY THE B SIDE OF THE TAPE

Suspect a defective amplifier, tape head, or dirty switches if deck B plays and tape deck A does not. Determine if deck A records. If the unit does not play or record, suspect that the tape head or amplifier circuits are defective. Determine if either or both channels are dead. If deck A will not play, but will record, check the switching and the tape head. If the same channel is dead in both tape decks, suspect the deck switching or amplifier circuits.

Clean all tape heads. Notice if the play head in deck A is stationary and not loose. Place a screwdriver blade in the high side of the play head or inject a signal at this terminal. If you hear a hum or tone from either test, suspect that an R/P tape head is defective. A good clean up might solve the playback problems of deck A.

NO FAST FORWARD IN DECK B

Check both fast forward and rewind of tape deck. If the rewind operates, but the fast forward doesn't, check the button assembly. Make sure that it locks into position. The fast-forward operation speeds up the forward motion with a fast-forward motor or an idler pulley and gears. Make sure that the idler wheel or gear is engaging take-up spindle (FIG. 10-12).

10-12 Make sure that the idler gear is engaging the take-up reel or turntable in the fast-forward mode. Clean the rotating surfaces with alcohol and a cloth.

On some models, the fast forward might operate from a higher speed motor. In others, the fast forward might operate entirely from a fast-forward motor (usually found in expensive units). Check the fast-forward motor drive belt. The drive belt might go to a regular motor or it might operate from a pulley on the capstan flywheel (FIG. 10-13). Clean the belt and drive surfaces with alcohol and a cloth.

If both fast forward and rewind are slow, check the motor belt drive and motor that drives both operations. A dry or gummed up motor might cause slow

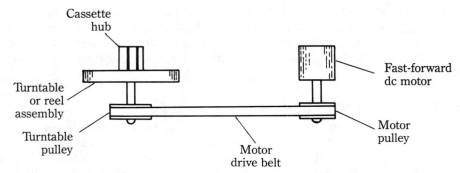

10-13   Suspect that the fast-forward drive belt is loose or has oil on it if it is running slow.

speeds. The dirty or oily motor drive belt might cause slow motion. Often when fast forward and rewind is slow, play is also slow. Clean all running idlers, belts, pulleys, and motors.

DOOR DOES NOT OPEN ON DECK B

The door might be jammed or a small lever release might be binding if the door will not open when no tape is loaded. If the door will not release, check for binding at the bottom of the door. Remove the front door panel, if possible. After removing front plastic panel, look for a small plastic or metal lever door catch.

Lift the catch with a small pointed screwdriver. If the door opens, the release lever is binding or not releasing the small holding lever. With the door open, work the stop/eject button. Notice if the button goes completely down to clear the release lever. Clean the levers and lubricate it with a drop of oil. Check for a bent eject lever (FIG. 10-14). Try to straighten any bent levers because they are difficult to obtain.

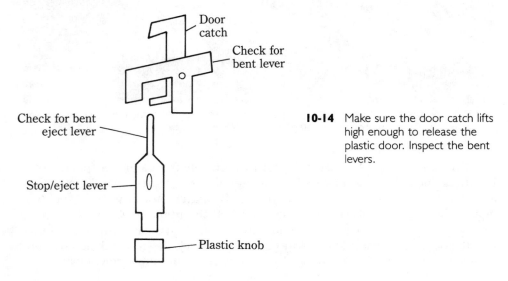

10-14   Make sure the door catch lifts high enough to release the plastic door. Inspect the bent levers.

The door catch might be bent out of line and not move from the plastic door piece. Notice if the front door is out of alignment. The door catch is under spring tension and should push the lever down to hold the plastic door assembly. If the catch is worn or slips off, bend it so that it will hook each time. When the front door is pushed in, the lever catch slips into the plastic notch so that door will not open. Clean excess oil or grease off of the metal door catch.

NO SOFT RELEASE ON DECK A

Both door soft release mechanisms operate in the same manner. When the eject/ stop button is pressed, the door pops out on cassette decks with no soft release. The door slowly eases out with decks that have soft-door release (FIG. 10-15).

10-15 Clean the soft-release gear if the door will not open or if it opens part way.

If the door only opens part way, either the door is jammed, the lever has broken teeth, or a small plastic gear is binding. Check to see how the deck number B soft release button and lever assembly works. Try to rotate the small gear with your fingers. Inspect the gear lever for binding teeth. Check the door for broken springs. The door should push open without any help. When the door pops out quickly with soft release features, check for broken teeth in the door lever or for an off-track door.

RECORD SAFETY LEVER DOES NOT RELEASE

Load the cassette for recording. Make sure the small plastic tabs on top of the cassette are in place. If the plastic piece is knocked out, check the recording cassette for one to keep or save. If not, place tape over the opening. In case the small lever will not let the cassette load, suspect that a record lever is broken or jammed (FIG. 10-16).

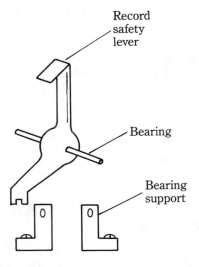

Record
safety
lever

Bearing

Bearing
support

10-16 Inspect the record safety lever if it will not move back or release. Check for foreign material.

Move the record lever back and forth with your fingers. If you can't remove the record lever, check for a frozen lever or bent side lever. The end of these levers might break off and let every cassette be recorded. If one of your favorite recordings is recorded over, suspect that a record lever is broken.

NO PAUSE

The pause control releases or moves the pinch roller from the tape and capstan to prevent rotation of tape. Although the motor might be operating, the capstan that turns the tape might be at a standstill (FIG. 10-17). Suspect a jammed pinch roller assembly, or a broken spring or lever when the pause button fails to pull the pinch roller assembly away from the capstan.

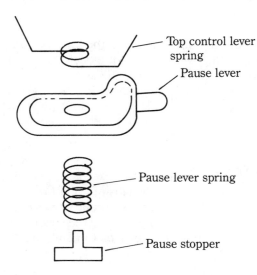

Top control lever
spring

Pause lever

Pause lever spring

Pause stopper

10-17 Check for a loose, off, or broken springs if the pause lever will not raise the pinch-roller assembly from the capstan.

Try to release pinch roller from the front loading door with the door removed. The pinch-roller bearing might be dry or gummed up and not pull away from the tape and capstan. A drop of oil on the hinged pinch-roller assembly might loosen the pinch-roller assembly. Do not spray oil inside this area because it will get all over the pinch roller and drive assembly. Check for a broken or loose control spring on the top, and a pause lever spring underneath the pause lever.

INTERMITTENT DUBBING

Intermittent dubbing might be caused by a dirty dubbing switch, mic, or mic connections. Determine if the mics operate in the regular recording mode. Often, a recorded tape cassette is duplicated from deck B onto a blank tape in deck A. Make sure that both R/P heads are clean in each deck.

Determine if tape deck A records and tape deck B plays back normally. If both decks are operating perfectly and dubbing is still intermittent, suspect a bad cassette or poor dub switching. Insert another cassette. If dubbing is still troublesome, clean the dub-control switch contacts.

POOR ERASE

The erase head must touch the tape in the record mode. Clean the erase head with alcohol and a cleaning stick. Notice if the erase head is stationary or if it swings up to touch the tape in the record mode. The erase head might not lock into position and touch the tape head.

The erase head is mounted ahead of the R/P head to erase any recording before the tape rotates in front of the R/P head. The erase head is excited with a dc voltage or a bias oscillator signal. Check for a low dc voltage at the erase head in the dc circuits. The isolation and voltage-dropping resistor might be open or the record switch section might be dirty. Take the voltage and resistance measurements at the bias oscillator transistor to determine if it is defective. Check for an open erase head winding with low-RX1 ohmmeter setting.

TAPE DECK A DOES NOT RECORD

Often, within dual-cassette tape decks, only one deck will record. The other deck only plays. Try another cassette before tearing into the cassette player. Remove the front-door cover and place the unit in play. This will bring the tape heads to where you can see them. Clean all heads, capstans, and pinch rollers with alcohol and a cleaning stick. Try recording on the new cassette.

Notice if the cassette plays normally. If the unit plays, but will not record, the amp section and R/P heads are normal. The bias oscillator and switching might be defective. Clean the function and record/play switches. Spray cleaning fluid into each switch and work the buttons back and forth to clean them.

Suspect a defective or dirty erase head if the recording is jumbled and two recordings run together. The unit is recording, but the erase circuits are not working. Check the erase head winding for open or shorted conditions. Make sure the erase head is touching the tape.

When the R/P head will play and not record, check the bias oscillator stage. Measure all voltages at the bias oscillator (FIG. 10-18). If a scope is handy, check the waveform at one side of the tape head or secondary winding of T1. Test the transistor in the circuit. Check the continuity of each winding. Sometimes improper soldering can be touched up with a soldering iron. Suspect a defective or dirty record switch or power supply with no voltage at the collector terminal of Q1.

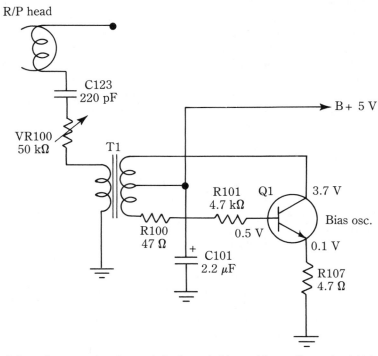

10-18   Check for a dirty erase head, or a defective switching or bias-oscillator circuit if deck A will not record.

TAPE DECK B DOES NOT PLAY

When one deck does not play, check the other. If tape deck B will not play and deck A does, the amplifier circuits are normal. Clean the playback head in deck B. Try another recorded cassette. Check the continuity of the playback head. Replace the head if it is open (FIG. 10-19). Look for broken cable wires around the tape head and clean the playback switch.

CASSETTE WILL NOT SEAT IN TAPE DECK B

Check for foreign material within the cassette holder when the cassette will not seat properly. Candy and gum wrappers will hold the cassette up enough that the door might not close. Notice if the play button is locked in. The mechanism might

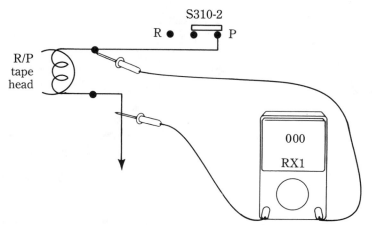

10-19   Check the continuity of the playback head in deck B if it has no audio.

be shifted up enough that the door doesn't shut. Inspect the loading mechanism for cracked plastic or a broken front door.

NO AUDIO ON TAPE DECK B

If a tape is rotating in deck B with no sound, suspect a defective tape head or amplifier circuit. Is the playback mode in deck A normal? If both decks have either the right or left stereo channel down, suspect that the amplifier circuits are bad. Pass the blade of a small screwdriver in front of each tape head. A thud noise will be heard with volume wide open. If the right channel has a low hum or is noisy with the volume wide open on only tape deck B, the tape head or the connecting wiring is open.

Touch the center terminal of the right-channel volume control with volume wide open and listen for a loud hum. If you can hear the hum, the amplifier circuits between tape head, preamp, and volume control are dead (FIG. 10-20). Before checking the circuits, clean the function and record/play switches. You can either signal trace the recorded signal with an external amp or signal tracer and take critical voltages on preamp IC1.

Signal trace by the numbers. Start at the output of the tape head with the signal tracer and cassette playing. Follow the right channel audio through the audio circuits until the music stops. If the music is found at the input terminal 1 (3) of Dolby IC2 and not at pin 6 (4), suspect that an IC is defective or that voltages are improper. Notice that these voltages are very low and should be taken with a DMM.

Suspect a leaky IC when the voltages are low at pins 5 and 12. Check all components on each terminal to ground to determine if IC302 is leaky. If the voltages are off on most all terminals and signal is coming in at pin 1 and not at pin 6, replace IC302. When the signal is on one side of C218 or C219 and not on the other, suspect that an electrolytic capacitor is open. These small 1-μF capacitors

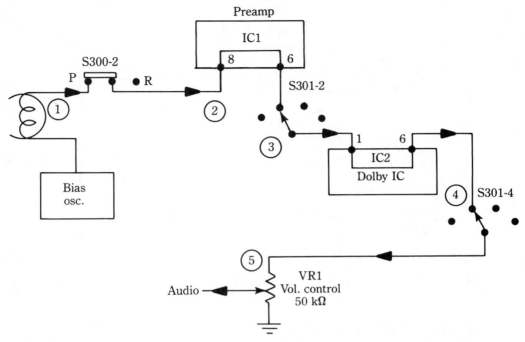

10-20 Check by the numbers with the signal tracer if no audio is on deck B.

might dry up and cause weak or dead audio. But a signal check on each terminal will determine if the capacitor is defective. Clip another electrolytic capacitor across the suspected one and notice if audio returns.

TAPE ROTATES/NO SOUND ON DECK A OR B

When both right and left channels are dead, suspect that a common component is tied to both channels. Check the voltage at the power supply tied to the power output IC or transistors. Determine if the sound is dead in both channels at the volume control. Use the screwdriver hum test or use a signal tracer. If no hum is heard at the volume control and the signal is traced to the volume control, suspect audio output problems.

Notice if both audio channels are fed into one large IC or if they are separate. Often, the output IC is defective when both channels are dead. In separate output ICs, suspect improper voltages from the power supply. Measure the power-supply voltage across the large filter capacitor. Next, check the supply voltage terminal at the suspected IC (FIG. 10-21).

Check the supply voltage at pin 1 of IC303. If the voltage is low (6 V) or if there is no voltage, suspect the low-voltage power supply. Remove pin 1 from the PC board by unsoldering the pin terminal. Now, take another measurement at the PC wiring. With pin 1 removed from the circuit, the voltage should rise above

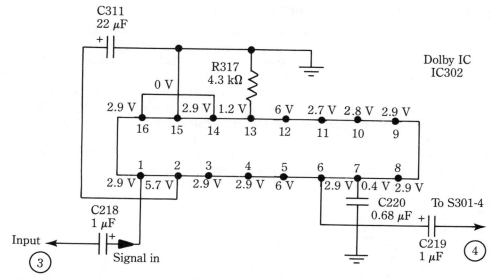

10-21 Especially check the supply voltage of the audio-output IC and all other terminals if no sound is on either deck A or B.

normal (10.5 V) if the power supply is good. Check the power-supply components with low or no supply voltage (FIG. 10-22). Replace IC303 if the power supply voltage is normal, the input signal is present at pins 5 and 8, and no signal is present at pins 2 and 11.

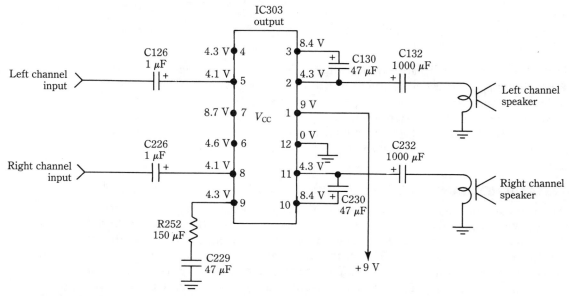

10-22 Check all voltages on each terminal of IC303. Low voltages might indicate that IC303 is leaky.

INTERMITTENT LEFT CHANNEL ON TAPE DECK A

Determine if the left channel plays back normally in both decks. If the left channel is intermittent on both tape decks, suspect an intermittent amplifier section. Both tape heads are eliminated. Intermittent audio is very difficult to locate and should be isolated in the front end or rear audio sections.

Signal trace the intermittent audio to the volume control by clipping to the top terminal with the left volume control turned down (FIG. 10-23). Play a recorded music cassette and listen for intermittent music. If the music is normal here, connect the audio signal tracer or amp to pin 5 of IC301. Rotate the volume to the normal position. Suspect VK104 and C126 if the sound becomes intermittent. Proceed to the audio output pin of IC301 (pin 2). If the sound cuts up and down, replace IC301.

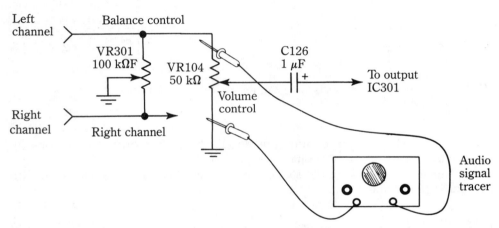

10-23 Signal trace the audio with an external amp or a signal tracer at the volume control to determine if the audio is intermittent in the front end.

Suspect poor PC board connections, C132, or IC301 if the left channel is intermittent at the speaker output terminals (FIG. 10-24). Do not overlook the possibility that the earphone-switching jack terminals might be defective. The audio signal can be signal traced right up to the speaker terminals with an outside signal tracer or external amp. Most audio intermittent problems are caused by poorly soldered connections, bad transistors, a bad IC5, or bad electrolytic coupling capacitors.

METAL (CRO) CASSETTE TAPE

The normal, chrome, ferrochrome, and other audio recording tapes are surface-coated with magnetic powder of iron oxide. Metal tapes have the surfaces covered with an unoxidizer alloy whose main ingredient is pure iron. Because base film is almost the same in all kinds of tape, the metal tape does not differ in regard to physical and mechanical properties. The metal tape excels in terms of electrical characteristics because of the special surface coating. The metal tape achieves superior fidelity in the recording and playback of music.

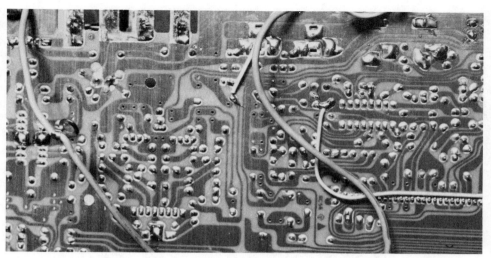

10-24 Poorly soldered board connections might cause intermittent sound.

The metal tape has excellent frequency responses at high output levels, both in the recording and playback modes. Metal tape also has high maximum output levels over the entire frequency range, a wide dynamic range at high frequency levels, superior sound to noise (SN) ratio at high frequency levels and an excellent distortion factor at high input levels.

Before recording on metal tape, make sure that the tape heads are clean. This is necessary to obtain maximum satisfaction from metal tape at high frequencies.

Suspect either a defective metal cassette, a dirty tape head, or improper voltage switching when high frequencies are distorted and not clear while playing the metal tape recording. Try another metal cassette and clean the tape heads. Check the CRO switch position, which is located on the front plastic cover. Make sure that the switch is on CRO and not normal bias when recording on a metal tape. Actually, the metal/CRO switch places higher voltage to the bias oscillator circuits (FIG. 10-25). R304 and R303 are paralleled in the B+ 5.5-V circuit.

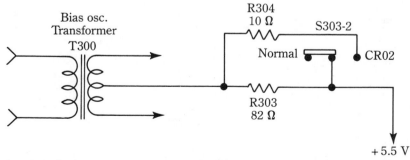

10-25 The CR02/normal switch is turned to CR02 when recording on a metal cassette tape. A higher dc voltage is applied to winding T310 of the bias-oscillator stage with S303-2.

BIAS CURRENT ADJUSTMENT

In many models, the current bias control should be adjusted according to the manufacturer's specifications. Remove the chassis and connect a 100-Ω resistor in series with the grounded side lead wire of the R/P tape head (FIG. 10-26). This can be done when repairs are made on the chassis. Connect a VOM or DMM across the 100-Ω resistor. Place the recorder in record mode. Adjust VR3 (bias control) on the meter for the correct μV rating from the manufacturer.

A mechanical, electrical, and electronic troubleshooting chart is found in TABLES 10-1 and 10-2.

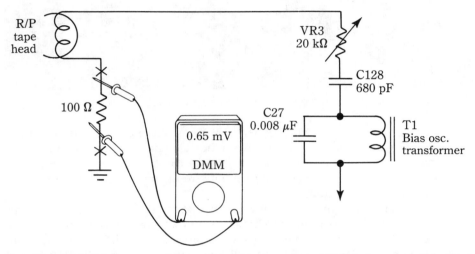

10-26   Check the bias current at the tape head by inserting a 100-Ω resistor in series with the R/P head terminal to ground. Adjust VR3 according to manufacturer's specifications.

Chart 10-1. Mechanical Troubleshooting Chart

Symptom	Trouble	Repairs
No recording or playback	Motor not rotating	Open motor. Test with ohmmeter. Open series resistor or diode. Defective lead or on/off switch. Frozen motor shaft. Dead or loose battery. Defective power plug.
	Motor rotates	Capstan rotates with oil or dust on pressure roller. Worn pressure shaft.
	Capstan does not rotate	Dry or gummed up capstan shaft. Drive belt out of position. Loose screw on motor pulley.
	Take-up reel or turntable does not rotate	Insufficient pressure on take-up pulley. Slippage of idler wheel.
	Twisted tape	Pressure roller and capstan not parallel. Record/playback head out of line.
Defective fast forward	Take-up reel or turntable slow rotation.	Oil on drive pulley. Oil on drive belt. Insufficient belt tension. Insufficient pressure at drive pulley. Worn take-up reel shaft. Defective tape counter.
Defective rewinding	Supply hub or reel too slow	Oil on drive pulley. Oil on belt. Oil on rewind idler pulley. Insufficient pressure of rewind idler. Worn supply turntable or reel shaft. Defective tape counter.

Table 10-2. Electrical Troubleshooting Chart

Defective Recoding on tape deck 1 or A	No recording bias	Check supply voltage on bias oscillator. Defective R/P switch. Defective leaf or on/off switch. Defective oscillator coil—check with ohmmeter. Defective R/P switch.
	Bias OK, but no recording	Level meter is operating—no record. Dirty R/P head. Defective R/P head. Dead no level meter—defective mic. Defective input jack. Defective R/P switch. Defective amp. Defective level meter. Defective meter diode or diodes.
	Defective erasing	Open or shorted oscillator coil. Defective bias transistor. Open or shorted erase head—check with ohmmeter. Defective capacitors in tank circuit of bias oscillator. Dirty erase head surface. Defective oscillator circuit. Improper supply voltage to oscillator circuits.
Defective playback	No sound, B+ supply	Open primary winding of power transformer. Defective leaf or on/off switch. Shorted B+ supply. Defective diodes in power supply.
	Voltages on all transistors and ICs are OK	Touch volume control for hum pickup—open R/P head or dirty switch. Touching volume control does not produce hum—defective speaker. Defective earphone jack. Bad speaker leads. Poor and dirty switch. Defective output transformer. Defective IC or audio transistors. Improper B+ voltage.
	No output, but noise and hum	Open R/P head.
	Defective tone	Shorted tone control leads. Defective tone control. Defective or dirty R/P head. Defective coupling capacitors.
	Low volume—abnormal voltage at transistors and ICs.	Check voltage on all transistors. Check voltage on all ICs. Defective coupling capacitors. Defective bias resistors. Defective small coupling capacitors. Defective large output coupling capacitor.
	Low volume output	Dirty R/P head. Defective IC output. Defective output transistors. Check voltages with normal channel.

Chapter 11

VCR repairs you can make

*A*lthough you need VCR service technician to make critical VCR repairs, you might save yourself some money by doing many of the common service problems. The minimum VCR service charge at most service centers is around $69.95. However, major service repairs, such as head replacement, might cost up to $250.00. The precision VCR is a very expensive machine, but you can keep those repair costs down by doing routine maintenance yourself (FIG. 11-1).

Simple repairs, such as head cleanings, demagnetizing, and checking for cracked or slipping belts can be made by the novice or beginner. A build-up of tape oxide might in time damage several components. Minor service problems can be made by almost anyone who can pick up a screwdriver or a pair of pliers. Preventive maintenance might eliminate costly repairs and damaged recordings. Knowing how your machine operates and how to check the VCR when it does not perform can save money.

BETA OR VHS

Basically, two different VCR systems are heavily used today. The Beta machine might have a better picture quality (your eyes might not see it) and was the first VCR placed upon the market. The Beta format has less track guides or spindles for the tape to go around, recording faster than the VHS machines. Of course, the Beta VCR will not record for long periods of time compared to the VHS system.

VHS video cassette recorders will record or play up to 6 hours on one cassette. VHS has outsold the Beta machine 5 to 1 and only a few manufacturers make the Beta machine, but many different manufacturers turn out VHS VCRs. Some people say the VHS system has less flutter. Both systems have good recording formats and you will find that more prerecorded movies are sold in the VHS format.

Because the Beta and VHS systems are widely found in most homes, the 8-mm VCR has just arrived on the home front. Tomorrow, the narrow 8-mm format

11-1   One of the first video tape RCA recorders to hit the market. Routine cleaning and preventive maintenance could prevent a costly repair.

might offer stiff competition to the Beta and VHS recorders. The 8-mm recorders are quite small compared to the VHS machines.

THE VIDEO CASSETTE

Remember, the VHS and Beta cassettes will not interchange (FIG. 11-2). The Beta cassette is small in size, compared to the VHS. Actually, the whole Beta cabinet is smaller and lighter than the VHS machine. Today, even the VHS recorder is much smaller and lighter than the first VHS models that were introduced for home use.

11-2   The VHS cassette is a little larger than the Beta cassette. Notice the take up and supply reels at the bottom of the cassette.

The video cassette has a take-up and supply reel inside the cassette plastic container. These reels fit down over the respective spindle assemblies inside the VCR. Load the cassette with the label up. Insert the hinged door into the recorder. Be sure to rewind each cassette after playing or recording before you store it.

The cassette should be stored vertically in a cabinet. Place the cassette in a dust-proof box or container. Try to avoid excessive moisture. Do not leave the tapes close to a humidifier and expect to use them right away. Keep the cassettes away from large stereo speakers and do not leave them on top of the machine. If the video recorder is mounted on top of the TV, remove the cassette each time and store it. A TV set with strong stereo magnetic speakers might wipe out your favorite recording. Be careful when demagnetizing the VCR tape head and keep the demagnetizing tool away from the various cassette recordings.

When purchasing a new movie type recording, let it warm up to room temperature before attempting to play. Most VCR machines have a dew indicator which might prevent playing or recording. In some models, if the machine does not operate for ten minutes, the recording might be out of sync until the machine warms up. In the meantime, you have turned every control on both VCR and TV set.

A defective tape might jam the VCR mechanism and result in a very expensive repair. Do not play a cassette that has a broken case or hinged door. Like the small cassette recorders, the cassette might have a dry plastic bearing that would slow down when recording or playing (FIG. 11-3). If a liquid is spilled inside the tape area, the cassette might slip, slow, and place liquid on tape heads and spindles; this would require a good clean up before damage occurs to the recorder. Do not try to repair a broken tape with a piece of adhesive tape because this action might damage the video heads.

11-3 Inspect the cassette for broken or cracked body parts. Slow speeds might result from a defective cassette.

Pulled or unraveled tape could occur if the take-up reel stops or slows down like any cassette recorder. The tangled tape might get inside the various spindles and gears, and require expensive repairs. The top cover must be removed from the front-loading VCR if the cassette will not pop up or cannot be pulled out after tape problems. Often, the top-loaded cassette can be removed by removing the top-loaded front panel. If the top-loading mechanism will not release, remove the top panel cover of the VCR. Remove all excess tape and clean the heads and spindles with head cleaner and accessories. Always choose a good cassette for recording. Try another cassette before suspecting trouble inside the VCR.

TOOLS NEEDED

Only a few tools are needed for basic VCR repair. You might have them in your workshop or service bench. If not, choose small flathead and Phillips screwdrivers to remove the top and bottom covers of the VCR. Use long-nose and side-cutter pliers to remove and replace the soldered components. The low-priced 35-W pencil soldering iron is ideal to solder various electronic components. Pick a good liquid cleaning cartridge, head cleaning kit, and cleaning spray. A pocket VOM or DMM is handy for voltage and resistance measurements.

HOW TO OPERATE THE MACHINE

The early VCRs have top cassette loading, but most of the present recorders have front-loading facilities (FIG. 11-4). Simply insert the cassette and push down on the top of the loading platform in the top-loading units. The tape will automatically rotate when play or record buttons are pressed.

11-4 Most early VCRs loaded from the top. Here, a cleaning cassette is inserted to clean the tape heads.

With front-loading machines, insert the cassette with the label face up. Push the cassette with your fingers until it is automatically retracted and loaded. In some recorders, a cassette indicator will light and indicate that the cassette is loaded properly. Check the cassette if it will not load properly.

To unload the cassette, push the stop button. In top-loading machines, hit the eject button and the platform will pop up. In front-loading machines, press the eject button to open the loading slot. Remove the cassette (FIG. 11-5).

11-5 Today, most video recorders are loaded from the front area. Here, a cassette is inserted and the mechanism will pull the tape into the machine, over the cassette turntables.

TO PREVENT ACCIDENTAL ERASURE

Every video cassette comes with an erasure-prevention tab (FIG. 11-6). When the tab is removed, the cassette will not record. To re-record such a tape, just place a piece of tape over the gap. Sometimes the VCR will not record and the machine will not operate. First, check the opening at the back of the cassette. If the tab is removed, the cassette will not record. In many cases, small cassette players and recorders have been brought into the shop for repair with only the tab removed from the back of the cassette.

RECORDING

Turn on the VCR power switch and let the machine warm up for 5 to 10 minutes. Select the channel you wish to record with the VCR channel selector buttons.

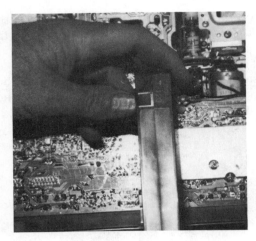

11-6 Each cassette has an erase-prevention tab at the back of each cassette. Remove the tab after you have made a recording that you wish to keep. Remember, you cannot record over any cassette if the tab has been removed.

Rotate the channel selector knob of the older models. Remember, the TV does not have to be turned on for station recording. However, you might want to see the channel you want to record. If so, set the video TV switch to the video position, turn on the TV set. Select channel 3 or 4: whichever channel has the weakest TV station so that it won't interfere with the VCR playback. The video channel switch is located in the rear of the recorder (FIG. 11-7).

11-7   Play the recording or cassette on channel 3 or 4 of the TV receiver. Set to channel 3 or 4 of the VCR for the best reception in your area.

Now load the cassette. Set the tape speed switch to the desired tape speed, SP, LP, or EP. The SP setting is best for music programs, and lengthy movies and sport programs are best recorded with the LP or EP switch positions. The LP/EP recordings can only be played on VHS-SP recorders.

Press the record button to begin recording. To keep commercials out of the recording, press the pause button to stop the recording. Press the pause button again to resume recording. Here the remote control comes in handy. Of course, if the recording is being made while you are gone or watching another TV channel, the commercials will be included in the program recording.

Press the stop button to stop the recording. In some models with preset time recordings, the recorder will automatically stop. All machines automatically stop when the end of cassette is reached. In fact, some machines will automatically rewind at the end of the tape. Press the rewind/reverse search button to rewind the tape.

PLAYING

Turn the power switch on and let the player warm up for 5 or 10 minutes if the room is cold or damp. Sometimes, if the recorder is damp and cold, the recording might be jumbled until the recorder warms up. Load a prerecorded tape or play the cassette to check out the cassette you have just recorded. Set the TV channel to 3 or 4 (whichever channel that the video switch set in at the back of the recorder). Press the play button. If noise or wavy lines appear, check the tracking adjustment knob. This should be left in center position while recording and playing modes. Now press the stop button. Press the (rewind/review Search) button to rewind the tape. Always rewind the tape after it reaches the end. Store cassettes in cassette containers or in a cabinet for safekeeping.

RECORDING ONE PROGRAM WHILE WATCHING ANOTHER

It always seems that the best TV programs are pitted against one another. Most VCRs will record your favorite program while you watch another. Select the VCR channel you want to record and start recording. In manually tuned TVs, make sure the program is tuned in properly. In electronically tuned receivers, the stations are tuned in automatically with the push of a button. Set the video/TV switch to the TV position and select the TV channel you want to watch, using the TV channel selector. If you want to check the picture during recording, set the video/TV switch to the video position. Then, set the TV channel selector to the video channel (3 or 4).

FEATURES AND FUNCTIONS

A good recording can be made when you know how and why each function switch works on the video recorder. Many times the VCR is brought in for repair when the operator did not know correctly how the VCR operates. Go over the record and play modes several times. Read the VCR instruction book thoroughly before attempting to make a recording. These machines are not complicated; most children operate them with ease. Here are the main features and functions of a Mitsubishi HS-328 UR model, to illustrate what each feature does:

Cassette loading slot   Cassette tape is inserted into this slot for loading (the recorder can be loaded manually or the cassette can be pulled in on some front-loaded machines).

Video/TV switch   When set to the TV position, the TV will receive off-the-air programs normally. When turned to video position, off-the-air programs selected by the VCR's built-in tuners are viewed on the TV. Set the video/TV switch to the TV position when recording one program while watching another.

Power switch   Press this switch to turn on power to the recorder.

Eject button   Press to remove the cassette.

Remote-control sensor   This is only found on recorders with remote-control operation. The sensor receives the infrared light from the remote-control transmitter.

Record button   Press to begin recording, both video and audio.

Start-time button   Press to select the start time for one-touch recording (OTR).

Record-time button   Press to record for 30-minute time intervals up to four hours.

Operation-mode display   Indicates operational mode.

Counter/timer display   This digital display covers operations in the counter/timer and program modes.

Channel indicator   The number corresponds to the selected channel.

Rewind/Review search button   Press to rapidly rewind the tape or to play the tape at high speed in the reverse direction.

Fast-forward search button   Press to rapidly wind the tape in playback or to play at high speed in the forward direction to search for a favorite recording.

Play button   Press to play a previously recorded tape.

Stop button   Press to stop all tape functions.

Pause button   When pressed during recording, it stops the movement of the tape. When pressed during play, a still picture will be seen, usually without sound. Press again to continue tape movement.

Channel selector   Selects the correct TV channel for recording.

Counter memory button   When pressed, the tape will fast rewind until the counter display indicates "0000" and will automatically stop at this point.

Counter reset button   Press to reset the counter to zero.

Counter/timer button   Selects either the counter or present time mode of the counter/timer display.

HOW TO CONNECT THE VCR TO THE TV

You might find 300- to 75-Ω antenna connections on the back side of the video recorder. Often, the round cable (75 Ω) is found at the VHS in and out connections. If your system uses round cable (75 Ω), plug it directly into the VHS antenna (in) socket. The output lead (comes with the recorder) is 75 Ω. Only one round cable might be used for VHS and cable hookup (FIG. 11-8).

With a flat ribbon cable (300 Ω) from the antenna, a 75-Ω adapter or matching transformer must be placed on the antenna cable for a perfect match. A push-on

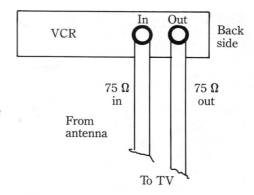

11-8 Remove the round antenna lead from the TV set and connect it to the 75-Ω VHF-input connections. Connect the 75-Ω output cable between the VCR and the antenna terminals of the TV.

VHS antenna adapter connects the 300-Ω flat cable to a 75-Ω fitting (FIG. 11-9). You might find a 300-Ω UHF and VHS in terminal at the older VHS recorders. Also, a small push-on matching transformer might be supplied with the recorder.

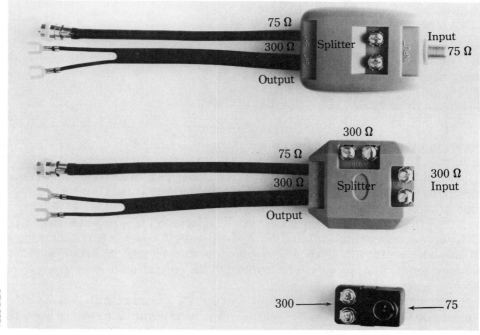

11-9 Many different fittings can be connected to the TV and VCR. Use the VHF/UHF splitter in A for a single 75-Ω input cable. The 300-Ω flat antenna lead is connected to the input of splitter B. A single push on the terminal connection with 300-Ω fitting is shown in C.

When both UHF and VHF stations are located in your system, they can be connected to the VCR with a matching transformer (FIG. 11-10). The signals from both VHF and UHF are fed from the booster system into a UHF-VHF splitter transformer. Here, only the VHF stations are recorded. The VHF signal is fed to the

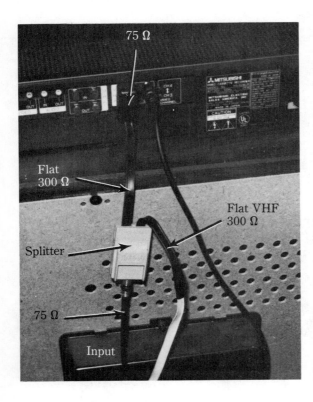

11-10 Here, a VHF/UHF splitter is tied between the VCR and TV. The round 75-Ω input cable connects to the splitter with 300-Ω VHF and UHF output fittings.

VCR through a matching push-on antenna terminal connection to the VHF in 75-Ω terminal. The 300-Ω UHF lead connects directly to the UHF terminal of the TV. If you want to record both UHF and VHF, the UHF signal from the splitter is connected to the in connection of the VCR, which is 300 Ω. Then, connect the 300-Ω out terminal to the UHF terminal of the TV.

The UHF and VHF antenna might be connected to the recorder (FIG. 11-11). When only one antenna lead or cable hookup is used, connect it as in drawing B. If only VHF is used in a given area, connect the antenna cable as in drawing C. All antenna input connections can be made with the correct splitter, matching transformer, and lead-in wire.

The output connections of the VCR to the TV set can be matched for either 75- or 300-Ω receiver antenna terminals. If the receiver-input terminals use a flat wire (300 Ω), place a matching transformer between the VCR cable (75 Ω) and antenna terminals (FIG. 11-12). Connect the VCR cable directly into the 75-Ω antenna terminal of the TV set with only a VHF connection. If the VHF and UHF outputs from the VCR come from one cable, use a VHF/UHF splitter at the receiver terminals. Here, the splitter might have either a 75- or 300-Ω output terminal connection, or one of each. Because many different antenna connections are available, select the right components for your VCR.

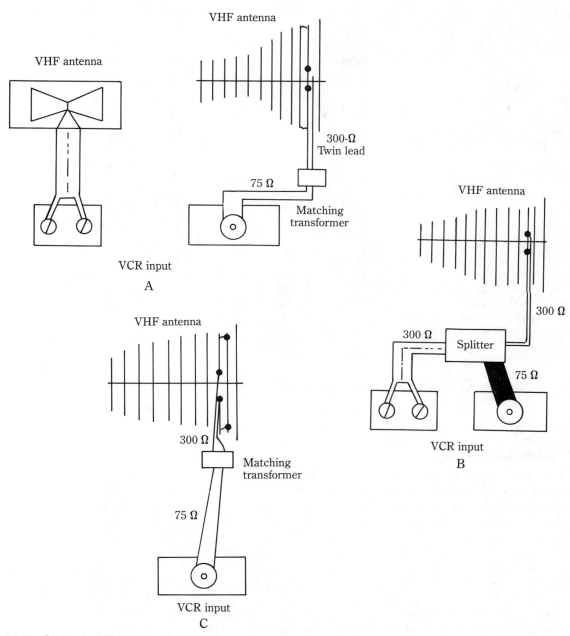

11-11 Both the UHF and VHF single lead-ins are connected to the VCR terminals. Only one lead in is connected with a VHF/UHF splitter in B. A single 75-Ω cable connects directly to the VHF input terminal of the VCR in C.

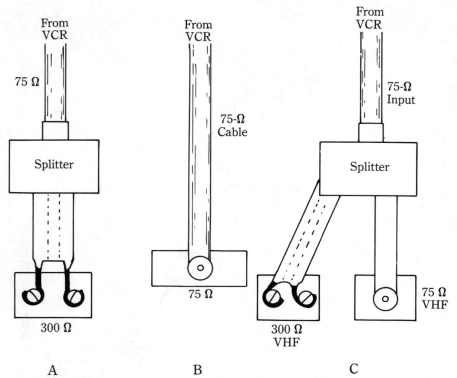

11-12 The back side TV antenna terminals might be 300 or 75 Ω. The VCR cable can be connected as shown in A, B, or C.

HEAD CLEANING

The dirty tape head might cause a loss of picture and sound. Excessive noise lines in the picture might be caused by a dirty tape head. When the tape head gap is closed with tape oxide, no picture or a partial loss of picture might be present. If the TV picture is clear, but during playback the picture is not clear, distorted, or snowy, expect that the tape head is dirty. The video heads might have a deposit of dirt or oxide built up, which would produce a poor picture.

To keep the VCR in tip-top shape, wipe the dirt and dust from the top of the machine every week. It's very easy for dust and dirt to filter into the top-loading machines. After recording or playing cassettes back 10 or 15 times, clean the tape head with a cassette cleaner. The cassette head cleaner looks like an ordinary cassette (FIG. 11-13). Some of these cassette cleaners will only clean the tape head and nothing else. Choose a cassette cleaner that cleans the erase head, tape guides, audio head, capstan, pinch roller, and video head, all at once.

The cleaning cassette might work dry or wet with liquid applied through a slot opening. Simply apply 4 or 5 drops of the cleaning solution to the exposed

11-13 The video-cassette head cleaner might look like the ordinary cassette. Headwipes™ or Texiswabs™ can be used to manually clean the tape heads. Head-cleaner spray can be used to keep oxide off the tape heads.

felt pad. Also, apply a few drops to the four window slots. Insert the cleaner into the VCR as you would a video tape. Push the play button to start the cleaning cycle. Let the tape run for 15 to 20 seconds, then eject the cleaning cassette.

A video head cleaning spray might be used intermittently to help keep those heads clean (FIG. 11-14). Hold the spray can 4 to 6″ from the head or other parts to clean. The force of the spray might clean dirt and dust from the tape heads or guide assembly. Use the extension tube for tight spots. In most cases, the top cover must be removed to get at the various heads and tape guides.

To clean the tape heads manually, the top cover must be removed from the VCR. The plastic front cover of the top-loading machine is held in place with several small bolts on top of the machine (FIG. 11-15). Remove the side screws from the VCR to slip off the top cover of the front-loading machine. You might have to remove the top loading platform or protective shield to get at the tape heads (FIG. 11-16). Usually, removing a few metal screws will release the metal shield.

Of course, the tape-head cleaning cassette should be followed up with a manual clean up after 48 or 60 hours operation (at least once every three months for the average viewer). A manual clean up consists of applying a cleaning pad and fluid to the various tape heads. Do not grasp the video tape head on the front surfaces with your fingers (FIG. 11-17). Oil or residue on the fingers might be transmitted to the tape head and tape.

11-14   A head cleaning spray can be used to clean dust oxide from the various tape heads. The long plastic nozzle can be stuck through the loading area to reach the tape heads.

11-15   The top cover must be removed to get at the tape heads in a top-loading recorder.

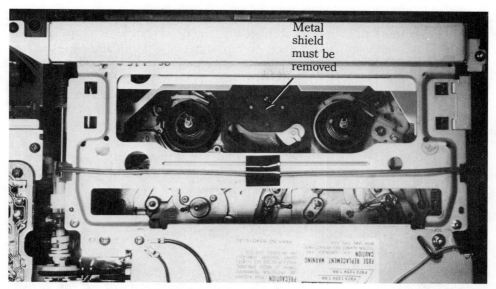

11-16 In this model, the metal shield bracket must be removed to manually clean the tape heads and guide assemblies.

11-17 Keep fingers off the tape-head face area to prevent oil or moisture, which could be applied to the tape. Hold the rotating video head at the top with the first two fingers while cleaning the curved surface.

Lightly apply cleaning fluid around the large video head instead of using an up and down motion. You might damage the tape head gaps with too much pressure. Hold the head with your fingers on top of the video head and wipe the area with your other hand. Look for a packing of brown oxide at the tape head gaps. Wipe all the other heads and moving parts with a pad and cleaning fluid. A professional cleaning product for all magnetic recording heads, called Headwipes™ contains a saturated pad. Simply tear open and pull out the wet pad, then clean the heads or tape-guide assemblies (FIG. 11-18).

11-18 Clean all tape heads and guide assemblies with commercial Headwipes™ or regular cleaning pads. Do not use cleaning sticks, such as those which are used to clean stereo cassette decks.

BEFORE CALLING FOR PROFESSIONAL SERVICE

When the VCR will not light, check the power cord, outlet, and on/off switch. On some models, the progress/record switch must be switched off. A shaking picture or no color might be caused by an improper setting of the TV or of the recorder channel selector. Readjust the fine tuning. Make sure that the channel (3 or 4) is set correctly.

Suspect a defective cassette or that the tab is removed at the back of the cassette if the VCR will not record. Try another cassette. Make sure that the video/TV switch is set to the TV position. Check the VCR channel selector if the channel is active channel, but the TV isn't being recorded.

Notice if the progress/record switch is on when the play button does not operate. Switch it off. Check to see if the tape is at the end if the fast forward doesn't work. Likewise, a no rewinding symptom could be caused if the tape is at the beginning. Readjust the tracking control if the picture is noisy or poor. Clean the VCR tape heads after the recorder has been used for a long time and the pictures are not clear after recording. Besides these minor adjustments, several visual and continuity measurements can be made before you cart the VCR off to professional service.

Demagnetizing the tape head

The magnetized tape head might cause a loss of color, weak and erratic pictures, noisy sound, and flagging symptoms. Flagging occurs at the top or bottom of the picture, bending the picture. You might notice the picture bending when a straight object, such as a house outline or building, is shown. Flagging and glitches might result from a dirty or magnetized head, or from a defective tape.

Pick up a regular VCR head demagnetizer for correct tape head clean up. Do not use the 8-track or cassette demagnetizer you have on hand. You might ruin the VCR tape heads. This same VCR demagnetizer can be used on all video, sound and erase heads in the recorder (FIG. 11-19). Demagnetize the tape head each time you remove the cover for good head clean up.

11-19 Besides the large video head, demagnetize the erase, sound, and separate function heads. Shut the demagnetizing instrument off after pulling it away several feet from the recorder.

Do not apply power to the VCR when cleaning or demagnetizing the tape head. Turn on the demagnetizing tool and bring it directly to each tape head. Keep the end of tool from touching the head surface. Slowly rotate the video head drum. After demagnetizing all heads, slowly pull demagnetizer tool away from VCR and shut off the power. If the tool is shut off close to the machine, you might magnetize several metal components on the VCR. This same method is used while demagnetizing a color TV screen.

Several cassettes will not record

Check the knock out tab at the rear of the cassettes when the VCR will not record. Place a layer of vinyl tape over the opening. Visually inspect the cassette to see if any of it is broken. Make sure the cassette is loaded with the label up. A defective cassette might cause slow or erratic speeds. Always purchase a good brand of tape for the best recordings.

Check the belts

Improper tape speed or no tape action might be caused with a broken or cracked belt. Inspect the belt for oil spots. A loose or stretched belt might produce slow speeds (FIG. 11-20). Clean all belts and pulleys with alcohol and a cloth. Slipping at the transport motor pulley might be cured with a coat of liquid rosin on the pulley area. Replace cracked, loose, or broken belts.

11-20 Make sure that the drive and capstan belts are not oily, loose, or broken. Clean the belts with alcohol and a cloth.

Visual inspection

Check the pressure roller, reels, and turntables for easy rotation. A sticky or worn pulley might cause improper speeds. Inspect the plastic tape guides for frozen or dry bearings. These plastic spindles should spin freely. A drop of light oil at the top of the plastic bearing might help. Wipe up all excess oil from the tape-drive areas. Replace any broken or cracked components.

Erratic operation

Check and clean the on/off switch when the machine erratically turns on or off. Notice if the pilot light is on. Inspect the plugs and contacts when a board or when portions of the PC board can be moved. Sometimes spraying contact cleaner inside the plug connection helps. Check for a broken wire or for a bad solder connection at the motor terminals if the motor movement is erratic. Burned or dirty relay contacts might cause erratic motor operation.

Burned components

Overheated resistors, transistors, and ICs might indicate that a defective component is nearby. Take accurate voltage and resistance measurements on the suspected IC and transistors with the DMM. A leaky IC or transistor might cause an overheated resistor to smoke. Check the schematic for correct resistor replacement after replacing the leaky component.

Tape head continuity

Check each tape head for low continuity or for an open head winding. A jumbled recording might result if the erase head isn't working properly. Check the continuity of the erase head. No sound might be caused by an open audio tape head. An open video tape head might cause the no-picture symptom when playing or recording.

No tape action

Inspect the belts and the motor if the tape doesn't rotate, but the tape indicator light still works. Listen for the motor rotation. Check the belt for rotation when the motor is rotating. If the motor does not rotate, check for faulty microswitches, relays, or a faulty motor. Measure the voltage applied to the transport motor (FIG. 11-21). Check for faulty key switches and play and record buttons.

Keeps blowing fuses

Suspect that diodes are shorted in the power supply if the fuse keeps opening after replacement. Look for a bridge rectifier or for four separate diodes in the power supply. Each one can be checked with the diode test of the DMM. Any low-resistance measurement in both directions indicates a shorted or leaky diode. Make sure that you replace the diode with another of the same amperage because some of these diodes vary from 3 to 10 A. Momentary transistor or IC flash-over leakage

11-21 A defective transport drive motor or pressure roller might prevent normal tape motion. Check the voltage at the transport motor terminals if the motor does not rotate.

might cause the fuse to open. If the fuse holds after replacement, do not worry what caused it. Improper dc voltage applied to the various sections might result from a leaky or open voltage regulator transistor (FIG. 11-22).

11-22 Suspect that a voltage-regulator power transistor is leaky if improper voltages are at the various circuits. Remove the regulator transistor and check the circuit.

Mechanical problems

Improper play or record functions might be caused with mechanical failure. Broken key and button functions might prevent manual loading or play/record functions. Mechanical levers or parts might jam and prevent tape operation. If the bent levers operate switches, the play or record might not work. Most mechanical operations can be visually inspected to find the defective tape action.

Squeaky noises

Any dry component might cause a squeaky noise in the recorder. Remove the top and bottom covers and load the cassette. Set the recorder play and check each moving component for a squeaky noise. Dry motor bearings, turntables, plastic guides, and pulleys cause most rotating noises (FIG. 11-23). Often, a drop of oil on the bearing area solves these noise problems. Replace motors that have worn bearings.

11-23 Dry pulleys, motors, gears, or turntable bearings might produce a rotating squeaking noise. A drop of oil might cure the squeaky bearing.

Sound problems

Erratic-, distorted-, or no-sound symptoms might be caused by defective ICs in the sound circuits (FIG. 11-24). Locate the sound section and take critical voltage and resistance measurements on the sound IC and transistors. A leaky IC might cause the problem. Check the audio tape head for excessive oxide, which could produce distortion. Take a continuity measurement of the sound tape head for a no-sound condition.

Poor PC board connections

When the VCR records or plays intermittently, suspect poor board connections. Intermittent snowy pictures or sometimes no picture might be caused by cracked PC wiring. If the machine is intermittent, try to isolate the intermittent section with the block diagram.

Many times the wiring will expand and crack around heavy parts or metal standoffs. If the unit is dropped in shipment or accidentally knocked off the TV or

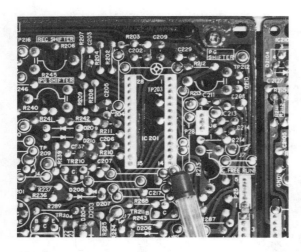

11-24 Distorted, weak, or no sound might be caused by the output IC. Take critical voltage and resistance measurements before replacing the suspected sound IC.

table, these boards might break or crack around heavy components. Do not overlook areas around metal insulators and braces.

Check the wiring with a magnifying glass because the eye might not see a fine crack in the narrow wiring. Sometimes applying solder over the entire section might help. If the cracked area is finally located, solder bare hookup wire over each broken wire. It's better to repair the cracked board than to wait months for a possible replacement.

TAKE IT TO THE EXPERT

After making the most obvious repairs and the recorder still does not function properly, take it to the dealer where the VCR was purchased, to a service depot, or to a local trained expert in VCR repair. Do not attempt to replace components in the critical recording and playback sections. Leave all required adjustments to the professional who has correct tension gauges and test equipment for critical adjustments. Do not turn screwdriver bias or controls without correct test equipment. Whenever possible, replace all parts with the manufacturer's part number. Do not dig into the chassis unless you know what you are doing. You might cause more damage than the original repair.

The VCR Troubleshooting Chart is found in TABLE 11-1.

Table 11-1. VCR Troubleshooting Chart

Symtoms	Trouble	Remedy
Dead-nothing	Power ac	Check outlet. Check power plug. Check fuse and power supply.
Blown fuse	Solenoid (main)	Shorted winding. Shorted diode.
No video	Bad tape	Play another cassette.
	Poor connections	Check cables and wiring.
	TV adjustments	Check for correct settings.
	Dirty tape heads	Clean.
	Defective tape head	Look close at head. Take open tests with ohmmeter.
No audio	TV adjustments	Doublecheck TV and VCR settings.
Video fair	Poor tracking	Readjust tracking control. Is sound normal on TV?
	Leaky or open ICs	Take voltage and resistance measurements.
Poor audio recording	Tracking	Adjust tracking controls
	Dirty tape heads	Clean audio ac head. Check bias oscillator circuits.
	Hum and buzz	Check audio cables and shields.
Poor color	Check tracking	Readjust tracking control.
	Check TV and VCR settings	Readjust.
	Dirty heads	Clean.
Erratic color	Color circuits	Press up and down on color board for loose components or connections.
	Poor socket connections	Clean and resolder connections.
Sound bars snowy pictures	Check tracking	Readjust tracking controls. Inspect video heads. Check video relay.
Noisy picture	Occurs at all speeds	Defective video head.
No video	Portion of picture clear	Check video head preamp, P/R switch, luminance circuits and rotary transformer. Check low-voltage power supply.
Wavy or picture wiggles	Tracking	Readjust tracking control.
	Dew circuits	Let VCR warm up 10 minutes.
	Dirty tape heads	Clean all heads.

Continued

Table 11-1 Continued

Symptoms	Trouble	Remedy
	Defective cassette	Insert a good cassette.
	TV adjustments	Readjust manual TV tuner and horizontal setting.
Jittery picture	Erratic supply reel	Replace worn and dirty brake band.
Picture dropout	Cassette	Insert another cassette.
	Tracking	Adjust tracking control.
	Head	Clean all tape heads.
No recording	Input cables	Check and refasten cables and plugs.
	No record or play	Check power-supply circuits.
	Cassette	Inspect for tab out at back of cassette. Place scotch or vinyl tape over opening.
	Dirty heads	Clean all tape heads.
	Jumbled recording	Check dirty erase head. Check for open head winding with ohmmeter.
	Defective cassette	Replace with new cassette.
No playback	Cables	Check all output cables and connections.
	Channel switch	Make sure TV is on same channel (3 or 4)
	Dirty heads	Clean heads.
Flickering color in playback	Video heads	Inspect video heads. Check automatic color-control circuits. Check video head preamps.
Erratic dropout	Cassette	Try another cassette.
	Dirty heads	Clean all heads.
	Tracking	Readjust tracking control.
	Tuner	Check for dirty tuner. Check tuner in TV set. Clean tuner contact.
	Tape head	Adjust tape head tension.
Tape will not load	Defective cassette	Try another cassette. Do not play broken or cracked cassette.
	Top loading	Check for binding levers or foreign material in holding area.
	Front loading	Check eject and front loading circuits. Check dc motor. Check foreign material like hair and gum wrappers in loading belt. Check for broken loading motor belt.
Cassette will not latch	Hold screw	Check for missing or loose screw.

Symptom	Component	Action
No tape motion	Power supply	Check fuses. Check power-supply voltages.
	Power switch	Check and clean. Replace if bad or broken.
	Belts	Inspect and replace broken or cracked belts.
	Bad transport motor	Check motor continuity with DMM. Replace motor with exact part number.
	Pressure roller	Inspect roller for worn areas. Check pressure on roller and capstan. Check for broken or missing roller spring.
	Reel problems	Inspect takeup and supply reels. Are they rotating properly?
	Solenoids and relays	See if solenoids energize. Take winding continuity. Clean switching contacts.
No record or fast forward	Relays and solenoids	Check for poor contacts. Check for energizing of relay.
	Micro switches	Are they open or closed. Replace if erratic and dirty.
	Belts	Check to see if loose, cracked, or broken. Off?
	Poor tape tension	Readjust.
	Motor	Check voltage to motor. Check for motor continuity. Replace motor.
No pause	Pause switch	Check to see if pressure roller is against capstan. Check tape brake. Check linkage.
Squeaky noises	Belts	Replace old or dry belts.
	Tape tension	Readjust—too tight.
	Gears, pulleys and turntables	Check for dry bearings.
	Guide spindles	Worn or dry-lubricate. Replace if excessively worn.
	Moving components	Notice if levers or wires are laying against moving parts.
HI FI sound	Problems on playback	Check audio heads on cylinder. Rotary transformer. Audio preamps. Audio switching relay.
Head wear	Excessive head wear	Check close up with microscope. Check with magnifying glass. Look for cracked or broken head.
Pulling or eating tape	Cassette	Replace with new one and try again.
	Mechanism	Check tape tension. Inspect turntable reels. Check reel brakes. Inspect defective pinch roller. Misaligned roller or tape guides.

Chapter **12**

Troubleshooting
stereo cassette decks

Stereo cassette decks are available in many sizes and shapes (FIG. 12-1). The home audio cassette deck might have a frequency response from 20 to 20,000 Hz, a high signal-to-noise ratio, and low wow and flutter. The cassette deck might include a dual or double deck and dubbing operation (FIG. 12-2).

The deluxe cassette deck might have twin auto-reverse tape transports with record and play, a 3-head transport mechanism with dual capstans for tape speed accuracy, a separate motor for fast forward and rewind, two auto reverse record/ play transports that let you record four sides nonstop or make two identical copies simultaneously, auto tape selector, blank skip, music search, reel tune counter, microphone inputs, recording level meters, and remote control.

The stereo cassette deck might also have an AM/FM-MPX stereo circuit. The deluxe radio is built to one side of the cassette tape deck (FIG. 12-3). You might find a phono turntable on top of some models. In older deluxe cassette decks, besides the AM/FM-MPX radio circuits, the older stereo 8-track player is located at the bottom (FIG. 12-4).

CASSETTE FEATURES

Many features require the best reproduction of tape possible, including wow and flutter below 0.08%, higher signal-to-noise ratio from 73 to 79 dB, and harmonic distortion as low as 0.6%. The deluxe cassette deck might have the following features:

Wow and flutter The variation of speed with a lower reading indicating the best rotation. The wow and flutter measurement of 0.10% is excellent for most cassette players, but 0.6% is best.

Auto reverse The tape deck will play both sides of the tape without actually changing the cassette. The mechanism changes direction and often has two capstan flywheels and tape heads.

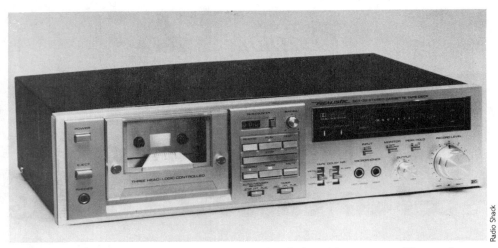

Radio Shack

12-1 The cassette deck might include both record and play with stereo amplifiers or line-out connections.

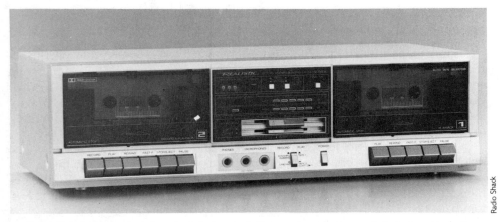

Radio Shack

12-2 This cassette deck has dual decks; deck B records and plays and deck A only plays.

Music search   The player will automatically stop at the beginning of a song or music when operated at fast-forward or rewind modes. Only the deluxe cassette players have this feature.

Frequency response   The lower and higher the frequency, the more ideal frequency response. The audible frequency range is 20 to 20,000 Hz. Often, women can hear above 15 kHz, but most men cannot hear above 10 kHz.

Noise reduction   The higher the signal-to-noise (SN) ratio, the better. Less hiss and better sound is ideal. Dolby C operates over a wider range, but Dolby B removes high-frequency hiss only.

12-3   The stereo cassette deck might include AM/FM-MPX receiver circuits.

12-4   In older deluxe cassette decks, you might find the MPX-radio receiver, recording meters, tuning meters, and 8-track tape player.

Auto tape selector   Bias and equalization is set automatically when the cassette is inserted into the tape deck. A different bias is needed when metal tapes are used.

Number of heads   The dual-cassette player has one deck with dual heads for record and play. In this case, play is found with one head. The record, play, and erase heads might be called 3-head transport. The Sony TC-WR670 has high-density hard permalloy heads.

Number of motors Most typical cassette players have one tape motor for normal speed. In the deluxe models, two tape motors might be used: one motor for accurate tape speed and another motor for fast forward and rewind. The dual deck might have 3 or 4 different motors. In the lower priced dual decks, one motor might rotate all functions (FIG. 12-5).

12-5 Most less-expensive decks use one motor for all functions, as in this Soundesign deck.

CAN'T OPEN DOOR

The front door might be warped or the door catch might not release when the eject/stop button is pressed. Do not try to force the door and break the plastic. If the front door cover can be removed, take it off to peer in at the door catch. Lift the catch mechanism with a small screwdriver. Replace the door if it is broken or warped and will not release. Repair the bent catch lever and check for loose or broken door hinges. Suspect that tape has spilled out if the door will not open and the cassette will not pop out (FIG. 12-6).

12-6 Remove the front plastic piece if the door will not open. Look for pulled tape wrapped around capstan and pinch roller, which would make the door and cassette difficult to remove.

CASSETTE WILL NOT LOAD

Peer inside the tape holder for foreign material, such as cigarettes and gum wrappers. Make sure that the record safety lever will release. Inspect the plastic holder for cracked or broken areas. Try another cassette. The cassette might be cracked or broken.

Make sure that the door closes properly. Is the recorder in the play mode? The mechanism might be high and will not let the cassette load. Usually, small items inside the cassette holder prevent proper cassette loading.

DEAD CASSETTE DECK

Go directly to the power supply when the dial lights and nothing works. If the unit operates on batteries and not ac, suspect that a power supply is defective. Check the low voltage at the power-output transistors or ICs. With no voltage, check the primary winding of the power transformer.

Measure the B+ voltage across the large filter capacitor. If the voltage is normal at this point, suspect that a resistor or voltage regulator is open. If a schematic is handy, check the power-supply circuits for a voltage regulator (FIG. 12-7). The voltage regulator might furnish voltage to the motor and amp circuits. Q11, R10, and D1 might open and prevent voltage at the 17- and 15.5-V sources. The open or leaky voltage regulator can be replaced with universal transistors.

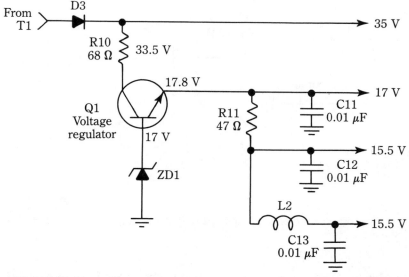

12-7 You might find more than one voltage-regulator circuit in the cassette-deck power supply.

Simply trace the voltage source from the cassette motor or amplifiers back to the power supply when a schematic is not available. The voltage regulator should be mounted on the power-supply board. Sometimes it is mounted on metal chassis with leads to the PC board. Do not overlook poorly soldered board connections. By taking accurate voltage and resistance measurements, the voltage source can be traced back to the full-wave rectifiers and the main filter capacitor.

KEEPS BLOWING FUSES

Suspect a blown fuse in larger cassette players when nothing lights. Often, the fuse is blown by a shorted silicon diode, filter capacitor, or leaky output solid-state device. If the fuse keeps blowing, suspect a leaky power output transistor or IC. Disconnect the output circuit from the power supply and cut a piece of foil or remove wires to suspected transistors.

Larger cassette players with higher power output might have four large transistors. Often, two are located in each output channel. Determine which channel is blowing the fuse by taking a low-ohm resistance measurement between the collector terminal and ground. After locating the two output transistors, test each one in the circuit for leakage. Remove them and test them for leakage out of the circuit. You might find one of the transistors shorted and the other open (FIG. 12-8).

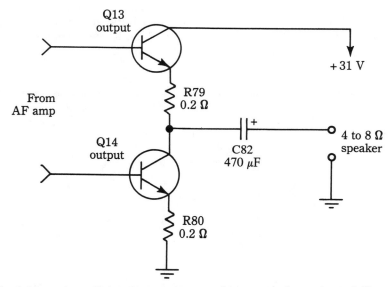

12-8 Check bias resistors if the power-output transistors are leaky or shorted. Test transistors out of the circuit.

While the transistors are out of the circuit, check for burned or open bias resistors. Usually, when a power-output transistor is shorted, the bias resistor opens. Also, while the two transistors are out of the circuit, check the driver transistor. Sometimes, the driver transistor becomes leaky and destroys the directly coupled power-output transistors. Most power-output transistors can be replaced with universal types. Leaky power-output ICs might also blow the fuse (FIG. 12-9).

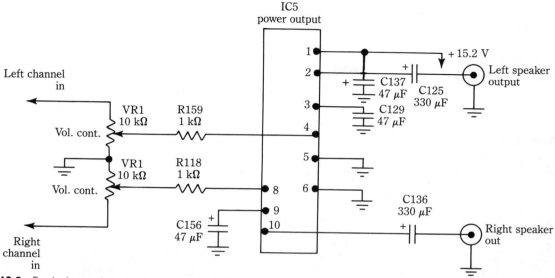

12-9 For leaky or shorted ICs, check the resistance at pin 1 to pin 6 at ground. If the measurements are under 100 Ω, the IC might be leaky.

STOPS AFTER A FEW SECONDS

When the tape deck keeps shutting off after only a few seconds, suspect that the automatic shut-off circuits are not working. In the units with automatic shutoff, a magnet is fastened to the end of a pulley on the counter assembly. Some models have a magnetic switch behind the magnet or IC. The magnet must keep rotating to keep the cassette player operating. When the magnet or tape stops, the magnetic switch or IC will shut down the operation (FIG. 12-10).

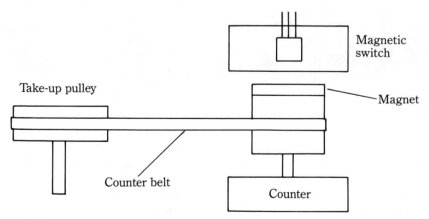

12-10 If the tape counter stops rotating with magnet attached, the magnetic switch shuts down the cassette player after a few seconds.

If the drive belt to the counter reel is broken, the cassette will start up and shut down at once. Look for a broken belt off of the counter pulley. Notice if the tape counter is rotating. If the belt rotates and the counter pulley and the unit shuts down, suspect a defective switch or IC. An IC is used in some Sharp models, but a magnetic switch is used in Sylvania tape players. The magnetic switch and IC are special components and must be obtained through the manufacturer or part depots.

SMOKING TRANSFORMER

Quickly pull the plug when the cassette player begins to smoke and groan when the unit is turned on. Check the primary winding of the transformer to see if it has opened. Check the B+ supply voltage or check the resistance across the large filter capacitor. A level below 100 Ω indicates a short circuit. Check the B+ at the output IC or transistor (FIG. 12-11) if it is easier to reach.

12-11   Suspect that silicon diodes, power-output transistors, or ICs are shorted if the power transformer runs too warm and smokes.

Check each silicon diode for an overheating transformer. Often, a dead short or high leakage of an output IC or transistor will make the power supply draw heavy current. Arcing and dimming lights when the ac cord is plugged into outlet indicates a heavy overload. In this case, the transformer is running hot.

Remove the ac secondary leads from the rectifier circuits to determine if the transformer has been damaged (FIG. 12-12). If additional secondary windings are found, remove them. Now, plug the unit into the wall outlet. If the lights still dim, the transformer makes a noise, or if it runs hot, the transformer must be replaced.

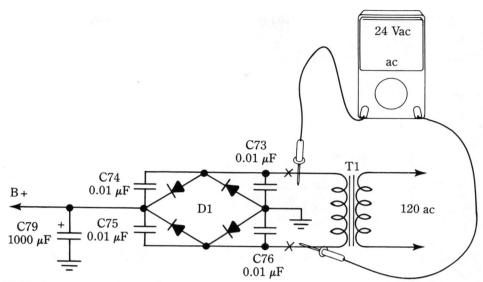

12-12 Remove the ac leads from the secondary of power transformer and measure the ac voltage. Replace the transformer if it runs hot after the load is removed.

The transformer is okay if it is cool. The transformer might be okay if the primary winding is not burned open. Replace the power transformer with exact part number or with a universal replacement if the original is defective.

NOISY OPERATION

Loud mechanical-type noises should be checked at once. When the noise is a hissy or fuzzy, suspect a noisy transistor or IC. Check for noise in each speaker. If you find the frying noise is found in only one channel with the volume turned down, the sound is originating in the audio output circuits.

Spray each transistor or IC with coolant and notice if the noise disappears or becomes louder. Spray several coats on each component before going to another. Sometimes, when the coolant hits the transistor or IC, the noise stops immediately (FIG. 12-13).

When the noise persists after you apply the coolant, try shorting each base of transistor or input of IC to ground with a 10-μF electrolytic capacitor. Start at the volume control and work toward the output or speaker. When the noise is cut down in volume or stops, you have found the circuit. Testing the suspected transistor does not help. Replace the suspect component. You might find a noisy ceramic bypass capacitor in the AF or driver circuits that is causing a frying noise.

NO REWIND OR FAST FORWARD

Usually the deluxe or expensive cassette deck has two motors. One motor is used for regular playback and the other speeds up to fast forward and rewind. Suspect a

12-13 The noisy IC might quit making noise if it is sprayed with coolant. This action will isolate the defective IC.

defective high-speed motor or circuits when the deck doesn't rewind or fast forward.

Like any motor, check the voltage that is applied to the motor terminals. Connect the voltmeter across the motor leads and push the fast-forward button. If it still doesn't move, try the rewind button. Suspect a dead or open motor when voltage is found at the motor terminals. Do not overlook the possibility that a diode or resistor might be open in series with the motor leads when no voltage is measured at the terminals. Look for a broken belt when you can hear the motor rotate. In older models, do not confuse the two motors because one might operate an 8-track deck (FIG. 12-14).

12-14 In this J. C. Penney cassette deck, the cassette player is mounted at the top and 8-track player is mounted at the bottom. Each has its own motor.

ERRATIC TAPE SPEED/UNEVEN PRESSURE ROLLER

Erratic speed could be caused by a loose motor drive belt, an oily belt, or a dry capstan. Check erratic speed with the torque cassette. If the torque is below 100 g, clean the motor belt, motor pulley, and capstan/flywheel. Then, again check the torque.

Uneven speed might be caused by a pinch or pressure roller that is not perfectly round or worn. Check along the rubber-bearing area for broken tape. Often, when the tape spills out and breaks, excess tape is wound around the pinch roller (FIG. 12-15). Remove the pressure roller if necessary and clean out the tape. Place a drop of light oil on the bearing. Wipe off any excess oil from the rubber area. Rotate the roller by hand to make sure it is free.

12-15 A dry or worn pinch roller, or one with tape wrapped around the bearing area might cause uneven speeds. Check the pinch-roller pressure with a tension gauge.

Check the pressure roller torque with a pressure gauge. Bend the roller pressure spring for more tension, if needed. The pressure roller helps pull the tape, with capstan, across the tape heads and feeds it on the take-up turntable. The pinch roller should run smooth and even. Replace the pinch roller if it is not perfectly round or if it is worn on the edges.

JUMBLED OR BAD RECORDING

No doubt, the erase head is not working when more than one recording is heard in playback. The erase head erases any previous recording before the tape passes the R/P head (FIG. 12-16). Place the deck in the record mode and do not record from microphones, radio, or dual deck. Let the cassette record for 10 minutes.

12-16 The erase head might have oxide packed in the gap area, preventing the erase of the previous recording.

Time it with the tape counter. Now, rewind the same cassette to the beginning and hit play. Notice if all previous recordings are removed.

If the tape still has several recordings on it, the erase circuits are not working. Clean the erase head and all heads with alcohol and a stick. Spray cleaning fluid in the function and R/P switch assembly. Now, start over and erase the tape again.

When the recording is still on the cassette after you clean the head, check the erase head for an open winding or a wire broken at the head terminals. Make sure that one side of the erase head is grounded. Measure the dc voltage across the erase head if it is meant to be excited by a dc voltage. Check the bias oscillator circuits with waveform bias.

REVERSE SIDE OF TAPE DOES NOT PLAY

In some tape decks, both sides of tape might be recorded and played back with a press of a button or in auto reverse. This is done with two capstan/flywheel tape heads and switching. Clean all tape heads. Check the wiring of the dead tape head. Clean the tape-head switching circuits. Measure the continuity of the tape head. Hold a screwdriver blade on the ungrounded side of the tape head and listen for a hum. Often the head or wires are defective because both heads use the same amplifier circuits (FIG. 12-17).

SINGLE-MOTOR FAST FORWARD

Fast-forward in a single-motor deck is done with mechanical idlers or gears to increase the speed. In some of these decks, another winding runs at a faster rate of speed. When the player is switched to fast forward, the normal motor wire is out of the circuit and the fast-forward winding is switched into action (FIG. 12-18).

With normal speed, the B+ voltage is fed to the main winding and through R21 and C51. In fast forward, the switch places the fast-forward winding to the B+ voltage and takes R21 and C51 out of the circuit. Here the motor speed is changed by switching the windings at the motor instead of via mechanical means.

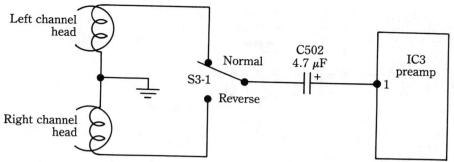

12-17 Check the reverse tape head and the head switch (S3-1) if the cassette player will not play other side of tape.

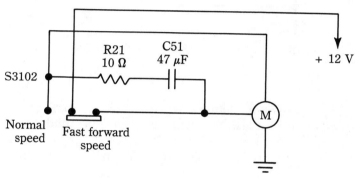

12-18 Some cassette decks have a single motor with normal and fast-forward speed.

NO RECORD ON LEFT CHANNEL

Check to see if the player will record on the right channel. Does the cassette play in the left channel? If so, the trouble might be a dirty R/P head. Sometimes oxide dust will pack into the small head gaps and cause poor recording or playback. Recheck the R/P head after you clean it to see if one of the gaps is packed shut. Next, clean the record/playback switch contacts (FIG. 12-19). Spray fluid into the switch area. Be careful not to let it drip on the moving surfaces. Make sure that the bias signal or voltage is reaching the left record head winding.

NO HIGH-SPEED DUBBING

High-speed dubbing is found in dual-deck cassette players. This feature enables you to duplicate a recorded tape cassette in tape deck B into a blank tape cassette placed in tape deck A. Some cassette players have high-speed dubbing, which is twice the speed of normal dubbing. Tapes dubbed at higher speeds are the same as those recorded at normal speed.

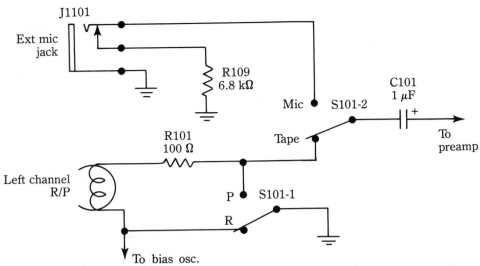

12-19   Check and clean S101-1 after cleaning the heads if the cassette deck will not record in the left channel.

Check the position of the dubbing switch. A dirty dubbing switch might prevent high-speed dubbing. Clean the switch with cleaning fluid. Usually, this cannot be done from the front of the cabinet. Remove the back cover and clean the switch by spraying into it while flipping the switch back and forth.

LINE-OUTPUT DECKS

In many of the earlier cassette decks, the cassette player/recorder schematic consisted of tape heads, transport mechanism, preamps, Dolby, and AF amp circuits. The AF amp stereo output fed into a line jack, which was cable-connected to a larger amplifier (FIG. 12-20). These cassette decks have all the features of a regular cassette deck, except for the power output audio stages and speaker terminals. The same troubles that exist in the line-output cassette players could be in any cassette player/recorder.

Line-output power circuits

The power supplies might be somewhat different from ordinary low-voltage supplies. You might find two separate secondary transformer windings; one going to a bridge and the other to a full-wave diode rectifier. The full-wave transformer winding is center-tapped to ground and fed into two separate silicon diodes. The output is filtered with a large filter capacitor (C21) and fed into a voltage-transistor regulator circuit. The regulated output voltages are fed to the input preamp and Dolby stages (FIG. 12-21).

The full-wave bridge-rectifier circuit has a separate winding, which is fed to a

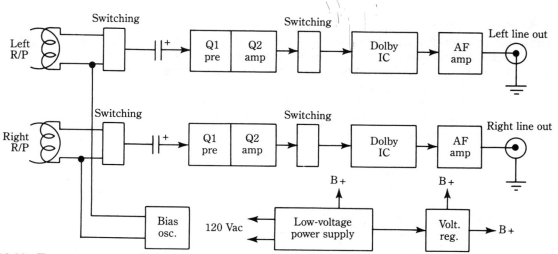

12-20 The stereo cassette deck might have line-output connections cabled to a high-powered amplifier.

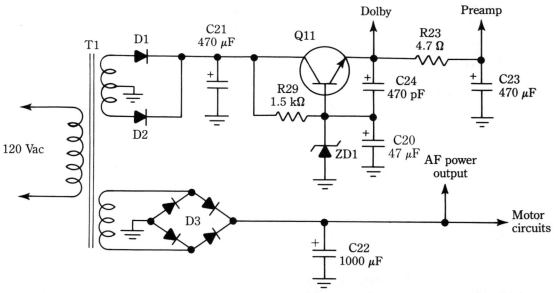

12-21 Two different secondary-power transformer windings might have voltage regulators feeding Dolby and preamp circuits, while the regular bridge supply furnishes voltage to AF, power-output, and motor circuits.

large filter capacitor (C22). The high filtered voltage operates the motor speed circuits and the collector terminals of the AF output line amplifiers. The bridge rectifier circuits might have pilot lamps in the transformer secondary and LED indicators within the B+ circuits.

NO AUTOMATIC STOP

The automatic stop feature might consist of a memory counter, reed switch, shut-off solenoid, and transistors. The reed switch is triggered with a magnet fastened on the back of the tape counter. The reed switch practically remains closed at all times because the magnet is operated at a fast rate of speed. When the counter stops or when the tape has reached the end, the reed switch opens and breaks the bias on Q1, which turns on transistor Q2, and energizes the shut-off solenoid that turns off power to the motor circuits (FIG. 12-22).

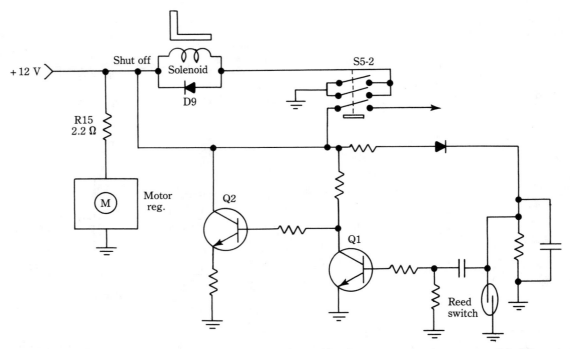

12-22   The automatic-stop circuits have a reed switch triggered by a magnet on the tape counter, Q1, Q2, and a solenoid with switches.

The automatic shut-off system might fail if the counter belt breaks or stops rotating. The shut-off system might fail if the reed contacts and switching contacts are poor, or if the transistors are open or leaky. Dirty motor-switching contacts or motor regulator circuits might prevent automatic shut off. The solenoid might shut down the tape rotation and leave the motor operating. Check all diodes, motor windings, and solid-state components within the motor-speed regulator circuits. If you can hear or see the solenoid energizing, you know the automatic circuits are functioning. Next, check the motor circuits.

BOTH STEREO CHANNELS DEAD

Always look for circuits that are common with both channels if the audio circuits are dead, intermittent, or noisy. A dual preamp IC and dual audio output IC are common to both channels. Do not overlook the power supply, which feeds voltage to both channels (FIG. 12-23).

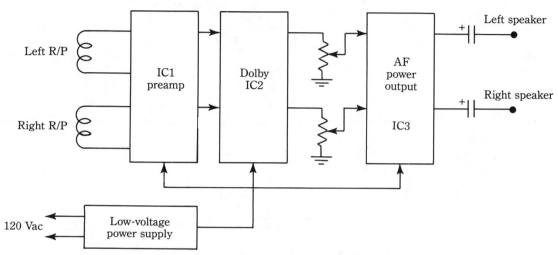

12-23   Check the power-supply voltage if both channels are dead.

Look for the most obvious circuit, the power supply. Do the pilot lights come on? If not, check the fuse and power supply. When the pilot lights are on, check the dual audio-output IC. Measure the dc voltage at the supply pin (V_{CC}). No voltage here might indicate that a B+ fuse is blown or that a power supply is defective.

Give the volume control the hum test on both channels or inject an audio signal. No sound indicates that a dual output IC is defective. If the output stages show hum or life, proceed to the dual preamp and Dolby IC circuits. Insert a cassette and signal trace audio from the heads to the preamp, through the Dolby IC to the AF circuits and the volume control.

DEAD LEFT CHANNEL

When either audio channel is dead, start at the audio output circuits. Most problems within the audio circuits are found in the audio-output and power-supply circuits. In transistor output stages, check the audio at the volume-control and output-transistor circuits. Poor speaker-terminal connections or an open speaker-coupling capacitor might cause a dead channel.

Quickly check the voltage at the collector terminals of the output transistors. Test Q12 and Q13 in the circuit (FIG. 12-24). If in doubt, remove one of them from the circuit for leakage tests. Remember, when one power-output transistor is leaky, the matching one might also be leaky or open. It's best to replace both.

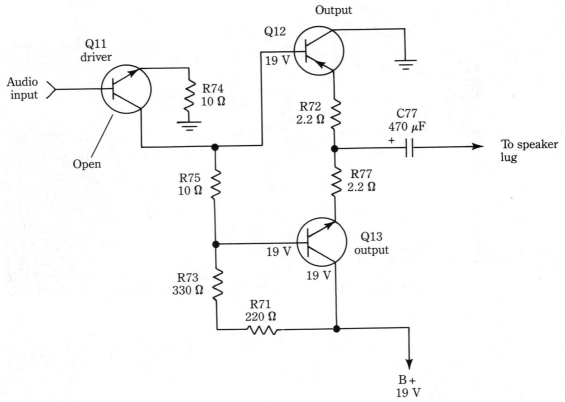

12-24 An open driver transistor might apply excess voltage on the push-pull power-output transistor with a dead audio symptom.

Do not overlook an open driver transistor. When Q11 opens, the supply voltage (19 V) might be found on the base of both Q12 and Q13. Make sure that all resistors are normal—especially R72 and R77 (2.2 Ω). One or both of these resistors might be open. When either output transistor is open, the voltages will change on all terminals of the transistors, except those that are tied to ground. If a schematic is not handy, compare all voltage and resistance measurements with the normal audio channel.

After replacing defective transistors, take comparable resistance readings on the collector, base, and emitter terminals to ground and compare them with the good channel. Make sure that all insulators are replaced under the power transistors. Take accurate voltage measurements on each terminal after audio channel is working and compare them.

DEAD RIGHT CHANNEL

Proceed directly to the dead right-channel output circuits and take critical voltage measurements on the transistors or ICs. Check the supply voltage terminal pin (V_{CC}) for the correct voltage. Measure the supply voltage at the metal body of the insulated transistors. Often, one output collector terminal is at ground and is not insulated away from heatsink. Be careful not to short the mounting screw to the B+ of the metal collector terminal (FIG. 12-25).

12-25 Small audio-output transistors are bolted to a chassis heatsink in a J.C. Penney 683-1773 cassette deck.

If the supply voltage is low at the power-output transistor pin, suspect a leaky output IC or defective power supply. Usually, a shorted IC will blow the fuse or cause the diodes to open or become leaky. Make sure that you are on the right channel when taking voltage and resistance tests. Remove supply pin 6 from the PC board to determine if IC1 is leaky or if the low supply voltage is in the power supply (FIG. 12-26). Most output transistors and ICs can be replaced with universal replacements.

INTERMITTENT LEFT CHANNEL

Intermittent audio is the most difficult problem to solve in the audio circuits. You might wait until the left channel quits or try to self-induce it to make the audio channels act up. If the left channel cuts out after a few minutes and stays out, try to isolate the intermittent sound ahead or behind the volume control. By isolating the front- and rear-end circuits, you can save a lot of service time. Take critical voltage measurements on the transistors and ICs in the dead channel. Don't be surprised when the dead channel comes to life when checking solid-state components.

Determine if the intermittent left channel has hum or frying noises when dead. Signal trace the signal up to the input terminal of the suspected IC with a signal tracer (FIG. 12-27). Sometimes, the voltage on all IC terminals might be quite normal with an intermittent IC. Check all components that are tied to the IC for

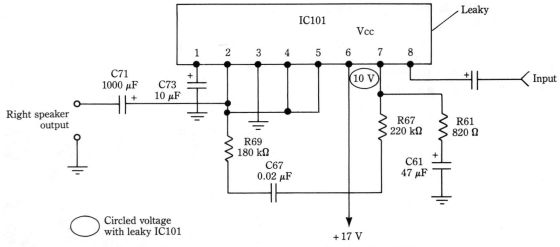

12-26 Often, the supply voltage pin 6 (V_{CC}) will be low if IC1 is leaky.

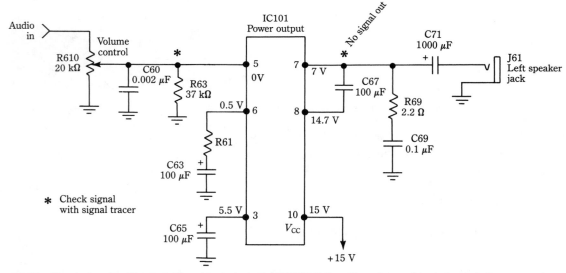

12-27 Check the signal in at pin 5 and out at pin 7 of IC101 if there is an intermittent channel. Suspect an intermittent IC if you hear frying noises or hum instead of the cassette recording.

correct resistance and voltage. Replace the output IC if a frying noise is heard in the noisy channel with normal input audio.

WEAK RIGHT CHANNEL

Clean the R/P tape head. Isolate the right-channel circuits from the left. Determine if the weak channel is before the volume control. Signal trace the weak signal from the tape head to the preamp and AF circuits in the right channel. If the signal is real weak, stop, and take voltage and resistance measurements. If the signal is stronger at the emitter terminal of Q20 than at the base, and very little voltage is at the collector, suspect a leaky transistor (FIG. 12-28).

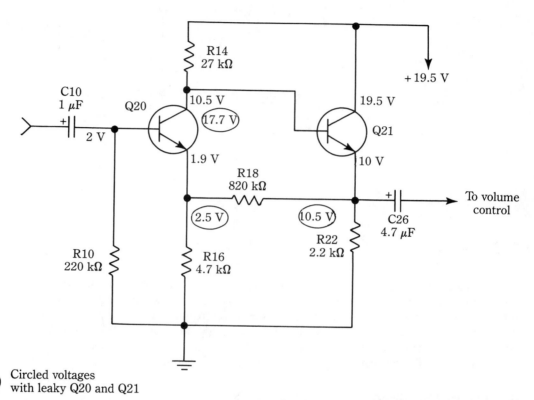

Circled voltages
with leaky Q20 and Q21

12-28 Weak channels might be caused by leaky directly coupled transistors in the AF circuits. With transistor and voltage tests, you can locate defective transistors.

When testing Q20 and Q21 in this circuit, both showed leakage. The collector terminal voltage was high (17.7 V), compared to the normal schematic voltage of 10.5 V. The emitter voltage on Q20 was high at 2.58 instead of 1.4 V. The resistance measurements from the base terminal of Q20 to ground was low at 870 Ω. Both transistors were removed from the circuit and found leaky. Remember, with directly coupled transistor circuits, a voltage change on one transistor will make the voltage read different on another transistor.

The audio can be signal traced from the R/P head back to the preamp, A, and volume-control circuits. If the signal is fairly normal going into the left input terminal of IC5 and very weak at pin 3, suspect that an IC is defective. Measure the supply voltage at pin 4. All voltages can be compared to the normal right preamp IC (FIG. 12-29).

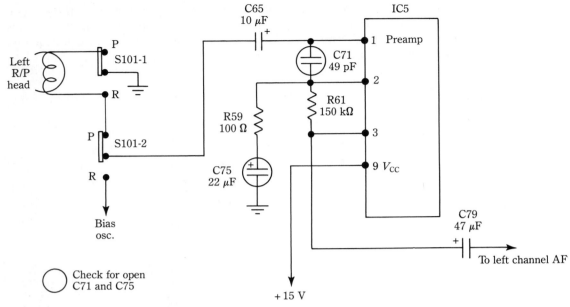

12-29   IC5 is defective if the signal at pin 3 is weak.

Check all components that are tied to the IC terminals in the right channel. If the voltages are normal, with input signal at pin 1, and very weak signal at pin 3, replace the suspected IC. Do not overlook small electrolytic capacitors tied to pin 2. An open C71 or C75 might cause a weak output signal at pin 3. Often, a resistance measurement is made to check for shorted components, but open capacitors and resistors might cause the same weak audio problem.

SPEAKERS AND SPEAKER CONNECTIONS

The cassette deck might have speaker plates, pushbutton terminals, jacks, and RCA phono jacks to connect speakers to the channel outputs (FIG. 12-30). The early models had multiple screw terminals or barrier strips. The line-output cassette recorders have earphone type jacks. Broken or frayed speaker wire at these connections might become dead, intermittent, and noisy.

Unless the cassette deck is one of the components that plugs into a large amp, most speakers are shelf-types. Up to three speakers might be found in the speaker cabinet (FIG. 12-31). Most of these extension speakers are under 10 W, but could go up to 50 W.

Interchange the channel speaker wires if one channel is dead or intermittent to eliminate a possible defective speaker. Remove the defective speaker from the cabinet and check it with an ohmmeter (FIG. 12-32). The voice coil might be blown open from excessive volume, which would result in no meter reading. Sometimes, the small speaker wire might break if it enters at the cabinet or at the speaker terminal.

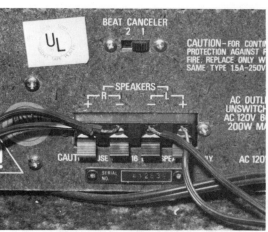

12-30 Pinch-type speaker connections are on the back of this Realistic 31-1995 model with a line fuse.

12-31 These outside shelf-type speakers, which are connected to the small cassette deck, can handle up to 50 W.

12-32   After isolating the defective speaker, check the voice-coil terminals with the RX1 position on an ohmmeter for the open winding.

Chapter **13**

Servicing compact cassette players

*T*he compact or rack cassette player might be in a single compact unit or stacked high with an amplifier, AM/FM-MPX radio, graphic equalizers, and record players. Often, the two large speakers sit beside the stereo or on a shelf (FIG. 13-1). Large high-powered stereo rack systems might have 110 to 120 W of power.

Some of the larger units have remote controls, high-speed dubbing, digital tuning, 5-bank equalizers, and 15-inch woofer speakers. The dual cassette deck might have continuous play mode, auto-reverse, and Dolby/BNR. The two-speed belt drive turntable might have a cue lever, an auto arm return, and a magnetic cartridge. The high-powered amp might have 120 W per channel, with 15-inch woofers, 4-inch midrange speakers, and 3-inch tweeters in large side-by-side enclosures. The matching rack has large glass doors (FIG. 13-2).

DUAL CASSETTE DECK PROBLEMS

In this Montgomery Ward JSA 39505 model, the number 1 deck (on the right) only plays, and deck number 2 records and plays cassettes (FIG. 13-3). You will find another record button in the row of buttons that records music. In most cassette decks, each unit has a separate motor.

One tape might play while the other is normal. Dead operation and speed problems cause the most problems. Slow speed might be caused by an oily belt, stretched belt, dry turntable bearing, dry capstan bearing, dirty motor pulley, worn idler, or a defective motor. A good cleaning of the heads, pinch-roller assembly, and all moving surfaces solves most speed problems.

Deck 2 no rotation/deck 1 normal

Try a cassette in each cassette deck to see if each deck is rotating. Within this GE dual-cassette deck, one motor rotates tapes in both decks (FIG. 13-4). Suspect a slipping belt, dry capstan bearing, or dry idler pulley when tape deck 2 will not rotate

13-1   The compact cassette player might have two large self-type speakers.

13-2   The rack compact system might have glass door enclosures.

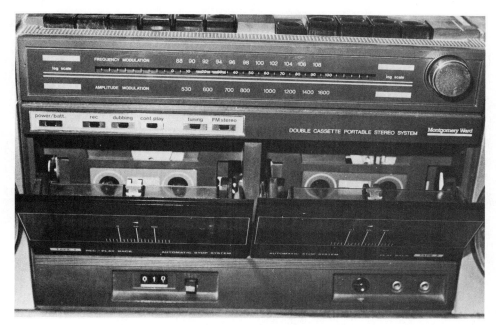

13-3 Montgomery Ward's JSA 39505 double-cassette stereo system.

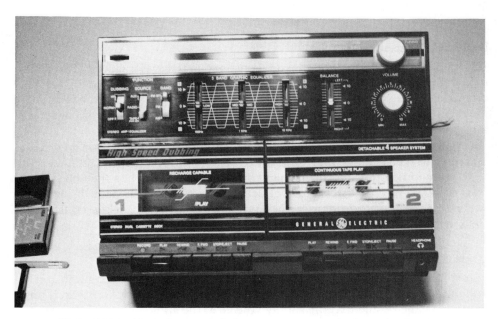

13-4 Within this dual cassette deck, one motor controls all rotating functions.

when deck 1 is normal. The motor circuits are normal when tape deck 1 is normal with one motor operating both decks.

In decks that have two different motors, suspect problems with the motor or circuits. You might find 3 or 4 motors in larger cassette decks. One motor is used in each deck for normal playback and a separate motor is used in each deck for fast forward and rewind. Measure the voltage at the defective motor terminal and compare it to the other deck. If the voltage is low, check the supply voltage or check for a change of resistance in the isolation resistors.

A defective motor-regulator circuit might cause the motor to run at different speeds (FIG. 13-5). Usually, with either a defective regulator circuit or motor, the motor runs very fast or slow. Clean the motor drive belts for speed problems.

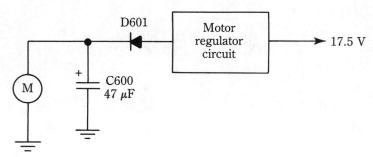

13-5 Always check the motor regulator system if the deck is operating too fast or slow.

Defective function switch

Several different function switches are contained in the cassette deck. The sliding-type function switch might be soldered directly to the PC board (FIG. 13-6). If it is dirty, clean this switch by spraying with the nozzle inside the switch area. Move the switch back and forth for clean up. You might have to spray several areas of the switch for proper clean up.

Rotating switches with several wafer sections have contacts that will oxidize and need to be cleaned. Spray each wafer section and rotate the switch back and forth. Sometimes, the long shafts might be bent or broken when the unit is tipped over or when heavy objects are dropped on function-switch shaft. Broken fiber wafers can be replaced. Replace the entire switch assembly if the contacts and shaft are broken.

Unusual dirty switch problems

Sometimes you might find a cassette deck with distorted or garbled audio and low volume. Go directly to the audio output circuits. A dirty function switch might cause the same problem. If the supply voltage to the audio output transistor or IC is at least half way below normal, a dirty function switch might cause the problem (FIG. 13-7). Measure the voltage on both sides of the radio/tape switch. If the voltage is normal at the switch and less than half at the switching terminal (SW715),

13-6 The spring-action sliding function switch is fastened directly on the PC board.

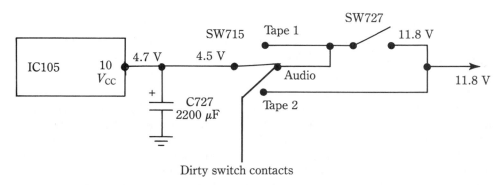

13-7 Weak and garbled audio can be caused by dirty switch contacts.

clean the dirty switch. To determine if the switch terminals are dirty, measure the resistance across the switch terminals with an ohmmeter; 2.5 to 3.5 Ω will cut the supply voltage in half.

Cassette loads/no play

If the cassette loads properly, but the tape does not rotate, the off/on leaf switch might be dirty, the motor might be defective, or belt might be off. Listen for the motor spinning when the play button is pressed. If the motor is operating, check the pause button. You might have pressed the pause button and forgot to release it. Check also for an oily or broken motor belt.

Measure the motor operating voltage at the motor terminals if motor will not rotate. Make sure you are on the right motor. Some of these cassette decks have a separate fast-forward and rewind motor, plus the regular tape motor (FIG. 13-8).

I3-8   The pen points at the play motor. The other motor is a fast-forward and rewind motor in this Technics RS-M 275X compact player.

Spin the motor belt by hand to determine if the motor is defective with the normal supply voltage. Check the tape switch or motor-regulator circuit with no or low voltage at the motor terminals.

Erratic play

Inspect and clean the motor drive belt, pulley, and capstan flywheel surface if the motor operates with erratic rotation. Rotate the flywheel by hand and notice if it appears dry or sluggish. Remove the flywheel support brace at the bottom and remove the flywheel (FIG. 13-9). Clean the capstan and the bearing. While the bearing is out, clean the bearing area with a cleaning stick and alcohol. A drop of light oil or phono lube on the bearing and end bearing provides adequate lubrication. Wipe up all excess grease or oil. Make sure belt is clean. If the rotation is still erratic, replace the defective drive motor.

Broken PC boards

Intermittent or erratic operation of audio or radio circuits might be caused by poor board connections. Check soldered connections where two different boards are joined together for cracked contacts (FIG. 13-10). On some PC boards, the speaker connections are joined by pushing the board into slotted areas where the PC board makes contact. Dirty board contacts might cause intermittent speaker reception.

Sometimes, the PC board is broken if the unit is dropped or if it is shipped brand new. The broken board might not be noticeable until the deck is several years old or is moved. Cracked boards can be repaired with bare or covered hookup wire on the longer connections. Never try to solder cracked wiring with just

13-9 Clean the motor drive belt and remove the capstan/flywheel for slow cassette speeds. Clean the flywheel bearing and lubricate it.

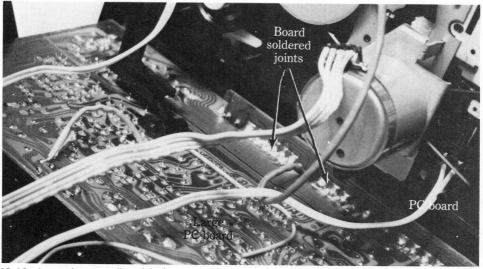

13-10 Intermittent audio might be caused by cracked solder contacts between two PC boards.

solder. It might hold for a few days, but will cause future problems. Scrape the enamel off of the PC wiring and tin the wiring with solder paste. Place a connecting wire across the cracked area. Make sure the wire you are connecting is on the same PC wire. Recheck connections with the RX1 range of the ohmmeter.

No radio recording

You might run into a cassette player that will record with the microphone and dub other recordings from cassette, but will not record from the radio circuits. Check for dirty switch and jack contacts. Does the level meter or indicator show that the radio is operating in the radio/record mode? If the record meter will not show signs of recording from radio, but if it does record with the microphone or dubbing of cassettes, signal trace the radio signal with an external amp or signal tracer to where signal is lost.

Signal trace the radio signal to the record/play switch or external microphone jack (FIG. 13-11). The stereo MPX tuner signal is applied to both left and right channels. Make sure that all switch contacts are clean. Sometimes, no radio recording is caused by improper grounding of the cassette player to the main radio chassis. Make sure that the two are properly grounded from the tuner to the cassette deck by taking an RX1 ohmmeter check.

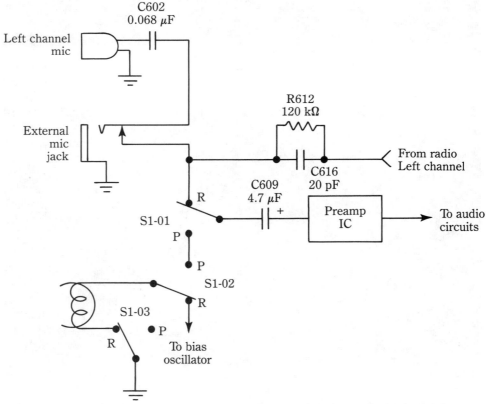

13-11 Signal trace the radio signal up to the microphone jack and preamp circuits that have no radio recording.

No record on the right channel

The left channel might record, but the right channel might be dead and not record. Clean the R/P heads and switches. Notice if the level meters are recording on both channels. If the VU meter or LEDs show no sign of recording, check the input signal from the tape head, switches to the preamp IC, or transistor stage. Does the left channel play with a recorded tape?

Isolate the no-record symptom to the play mode. If it plays in the left channel, but will not record, check the R/P record switch. Clean all record switches and input-jack terminals. Make sure that the recording switch lever is pushing the switch contacts to the record mode. Check the VU recording meters to see if it indicates recording in the left channel. The audio circuits are normal if the deck plays. Make sure that the right R/P head is clean.

Intermittent recording

If one channel records and the other does not, check for playback on the defective channel. Clean the heads and R/P switching. Determine if the erase head is erasing all information from both channels. Sometimes, you might hear a low recording on the defective channel. Make sure that proper beat bias is found at the intermittent recording head.

If the bias current is intermittent to the suspected tape head, the recording is intermittent. Check for bias current by inserting a 100-Ω resistor between the tape-head winding and take a voltage measurement. Notice if the meter hand changes or disappears if the channel is intermittent. If the bias current is intermittent, suspect a defective bias-oscillator circuit. Move and prod the various components within the bias-oscillator stage to locate a defective component. Do not overlook small bias (screwdriver or thumb) controls in series with each tape head. Sometimes, just touching the bias control will affect recording. Either clean the control with cleaning fluid or replace the control (FIG. 13-12).

The intermittent bias at the tape head in the record mode can be checked with the scope. The waveform might be high one minute and down to almost zero if it is intermittent.

Suspect that the bias-oscillator circuit is not oscillating if the bias is intermittent. After checking the bias at the adjustable resistor, resolder all terminals at the bias transformer (FIG. 13-13). Recheck the bias current or waveform. If it is still acting up, replace Q211 and both bypass capacitors in the base circuit of Q211. Besides transformer connections, these two capacitors might have caused intermittent bias in the R/P head. Q211 can be replaced with a universal transistor replacement. Do not overlook the possibility that C211 might be intermittent.

No VU meter movement in the right channel

The right and left recording level (VU meter) will move with the sound of the music. Notice the red area at the extreme right of each meter (FIG. 13-14). The

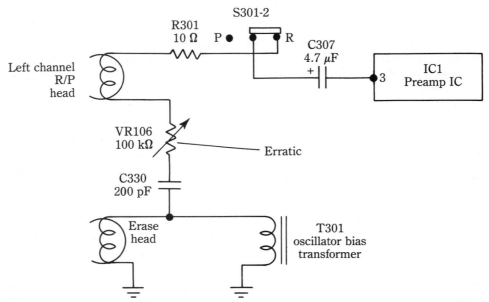

13-12   Intermittent recording in the left channel might be caused by thumb control VR106 from the bias-oscillator stage.

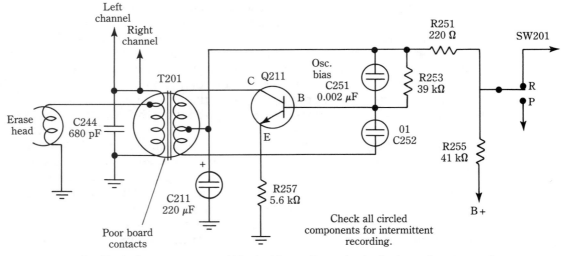

13-13   Check the components within the bias-oscillator circuits for intermittent recording.

recording-level controls are set to where the recording starts in the red area (0 to 4). Either VU meter might be used as an indicator if you are troubleshooting a no-record channel.

The VU meters are located in the circuit after the Dolby or recording-amp circuits. The audio signal is rectified by two diodes after the meter-drive amplifiers

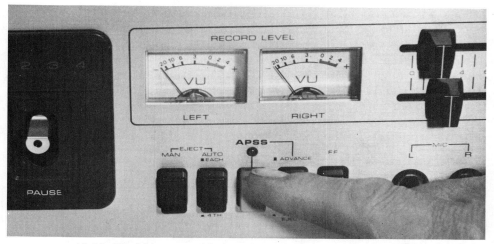

13-14 The VU meter might be open and not move while recording.

(FIG. 13-15). VR301 adjusts the signal for maximum reading on the meter. Check the meter resistance on the RX1 scale. If it is open, replace the meter. Suspect that D301 and D302 are open if the meter does not move. The audio signal can be signal traced at terminal 3 and 5 of IC301 through SW301-1 to the base of Q301. Check the voltages on Q301 and test Q301 in the circuit that has no meter movement.

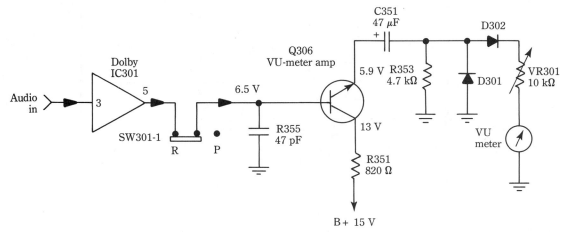

13-15 Suspect that the VU-meter amp (Q301, D301, D302, and VR301) is defective if the VU meter doesn't move while recording.

LED indicator problems

Very seldom does any one LED go bad and not light. Of course, the LED can be checked with the diode test of the DMM. The reading will be higher than that of regular silicon diodes. Suspect an LED meter amp is faulty if all of the diodes are out and will not respond to recording or playback modes. The audio signal can be traced right up to the input terminal of the meter amp.

If the LEDs are used in the record mode, the recording signal is taken after the preamp IC; in playback mode, the audio signal can be tapped after the audio-output IC through switch S703 (FIG. 13-16). The five LEDs are connected to pins 10, 11, 12, 13, and 14. The supply voltage is pin 4 (12 V). Measure the supply voltage and the pin voltage of each LED. If the voltage is fairly normal and if the audio signal is found on pin 3 suspect that IC701 is defective. Replace IC701 if the display is intermittent. Doublecheck all soldered pins of IC701.

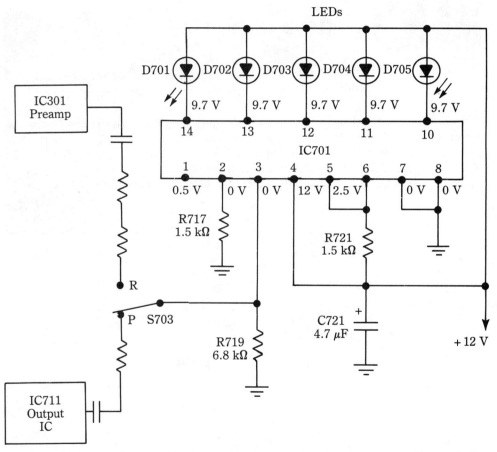

13-16 LED sound indicators might malfunction if the supply voltage is improper, if the switching is dirty, or if the LED amp IC is leaky.

Suspect the power-supply or voltage-regulator circuits if the voltage is low at pin 4. Some deluxe cassette decks have a voltage regulator for the preamp, bias oscillator, and LED amp circuits. A leaky IC701 might cause a low supply voltage.

Dead right speaker

Check for an open right-speaker fuse in larger cassette decks. Each fuse is protected with a small 1- to 3-A fuse (FIG. 13-17). Sometimes excessive volume will open the fuse. In directly coupled transistor-output circuits, the speaker is connected directly into the dc circuit. If the output circuits are balanced the voltage is zero at the speakers. If one of the transistors becomes leaky or opens, a balanced circuit is upset and applies dc voltage directly through the fuse to the speaker voice coil. The fuse will open to protect burning or damaging the voice coil of the larger speaker.

13-17 A dead right speaker could be caused by an open speaker fuse.

Dead left channel

When transistors are directly coupled to speakers, a fuse is used. F3 protects the speaker from voice-coil damage. You might find 5 or 6 transistors in one channel of the higher-powered output circuits. The audio output is balanced at F3 (FIG. 13-18).

If any transistors or other parts in the output circuits change in value, the circuit becomes unbalanced and a dc voltage is found at F3. Notice that a positive and negative 33 V is fed to the collector terminals of the two output transistors. If either transistor becomes leaky or open, you might find a positive or negative voltage at F3, which is applied to the speaker voice coil and opens F3. If F3 does not open, the voice coil is burned or frozen to the center magnet pole. Do not place a larger fuse in the clip or the speaker might be damaged.

If the fuse (F3) is open with a dead speaker, check the dc voltage at the fuse before inserting a new one. Of course, the fuse will open if a dc voltage is at the speaker terminals. Check for a dc voltage on both sides of the fuse.

If you find a dc voltage at the fuse terminal, check each transistor for open or leaky conditions. Both transistors might be leaky or one might be open and the other leaky (FIG. 13-19). Always replace both power-output transistors—even if

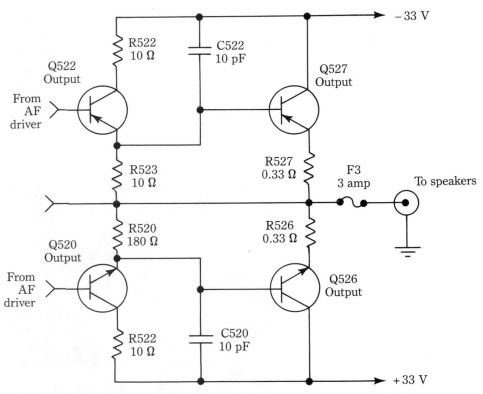

13-18 The balanced dead left channel might open F3 or damage a speaker if a leaky transistor applies a dc voltage to the 3-A fuse.

only one tests bad. Do not stop here. Test each transistor in the output circuits for open or leaky conditions.

A directly driven driver transistor might short and knock out both output transistors. Some of these high-powered amplifiers might have Darlington driven transistors in the output circuits. The Darlington transistor has a configuration of two separate transistors within one component (FIG. 13-20). It's best to replace them than to try to test them with the DMM or transistor tester.

Besides testing each transistor in and out of the circuit for leakage or open conditions, check each bias resistor and capacitor. Notice that the push-pull output-bias resistors (R527 and R526) have very low values (0.33 Ω). These resistors must be replaced with resistors that have the same values. Remove one lead of any bias resistor for accurate resistance. Often, these resistors are burned or overheated if the transistor becomes shorted or leaky.

The output circuit might become unbalanced if one of the power sources becomes low or is removed from the circuit. Check for an equal power-supply voltage at the B+ and B− sources. These voltages will be within 0.5 V of each other if they are normal.

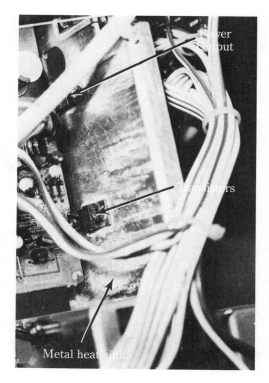

13-19 The low-output stereo transistor can be mounted on a large metal heatsink. Notice the white silicone grease around insulator and transistor.

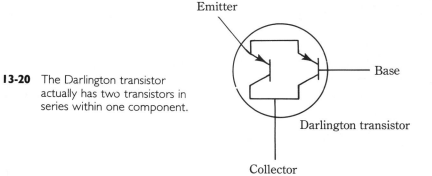

13-20 The Darlington transistor actually has two transistors in series within one component.

The leaky or open output transistor can be replaced with universal units if the originals are not available. Notice the small insulator found between the transistor and the heatsink. The heatsink might be the metal TV chassis, separate sheets of aluminum metal, or heavy ribbed heatsinks to distribute extra heat (FIG. 13-21). Always test the new replacement transistor before installing it. Apply silicon grease between the heatsink, insulator, and transistor. The grease heat-transfer comes in clear and white colors. Be careful because the white silicon grease can stain schematics and clothing.

13-21 Heavy TO-3 heatsinks with fins distribute heat from large power transistors.

After mounting the transistor, check the metal (collector) terminal of transistor to the metal heatsink for possible leakage. Do this test before connecting to the terminal leads. A good insulated transistor shows no leakage between the transistor and the heatsinks. The mounting bolts might be tightened with too much pressure and eat into the metal, which shorts out the transistor. Remove and replace the insulator if you find leakage. The insulator might short if a piece of melted solder ball, clipped bare wires, or debris is picked up with silicon grease. Keep both sides of the insulator transistor and heatsink free of any bench debris (FIG. 13-22). Some transistors have small round plastic insulators or washers that fit into each mounting screw.

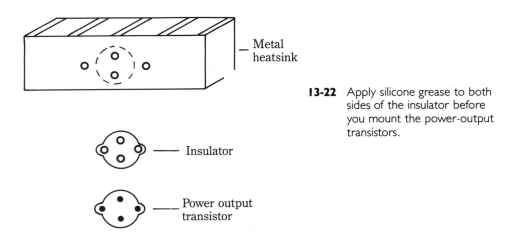

13-22 Apply silicone grease to both sides of the insulator before you mount the power-output transistors.

TO-3 and TO-220 mounting transistors use metal heatsinks. The metal body is usually the collector terminal. In some cases, the metal body or collector is at ground potential and is mounted directly to the metal heatsink without the insulator. Apply silicon grease between the transistor and the heatsink (FIG. 13-23).

13-23   Power transistors are mounted on large metal heatsinks that are bolted to the PC chassis.

Distortion in both channels

Suspect a common component if both channels are distorted. The leaky dual pre-amp IC, Dolby IC, or power-output IC might cause distortion in both channels. Often, improper voltages are found on the IC terminals, which indicates that an IC is leaky. Intermittent distortion might be caused by a defective dual IC with normal voltages. Replace the IC if it is defective.

Check for improper power-supply voltages if both channels are distorted. Remember, the low-voltage power supply is common to both stereo channels. Distortion might be caused in high-powered output-transistor circuits if a positive and negative voltage is applied. If one of the voltages is lower than the other, weak volume and distortion might develop.

Do not overlook a separate component, such as a power IC, that might be defective. Sometimes, both of these output ICs might have a flaw and become leaky or open at the same time. Unusual bases with small electrolytic capacitors within the speaker-output leads might become defective and distort (FIG. 13-24).

If the volume is increased, the distortion is greater. Sometimes, spraying coolant on ICs and transistors will make these components act up. Often, if a defective or dried up electrolytic capacitor is sprayed with coolant, the distortion is greater. Electrolytic coupling capacitors have a tendency to open or dry up and produce weak and distorted volume. An internal break of foil and lead results in intermittent audio at electrolytic coupling capacitors.

Distortion in the right channel

Most audio distortion in any channel is caused by transistors, ICs, electrolytic capacitors, and bias resistors. Switch speakers to determine if the speaker is distorted. Isolate the correct channel. Often, the audio transistors or ICs that are mounted on the right, looking at the front of cassette deck, are in the right channel.

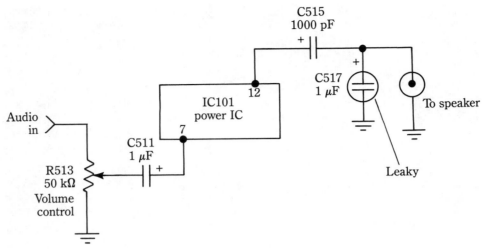

13-24 Both stereo channels might be dead if the same electrolytic capacitors became shorted in the speaker circuit.

Signal trace the distortion by going from base to base of each output transistor. Likewise, check the input and output terminal of each IC for distortion with signal tracer or external amp (FIG. 13-25). Check each terminal of a suspected electrolytic coupling capacitor. If the distortion begins in a given audio circuit, check the voltages and components in that circuit. Be careful not to touch two IC terminals together if taking voltage measurements.

Besides transistors in the audio circuits, look for a change in bias resistors. A leaky power-output transistor usually damages the corresponding emitter resistor.

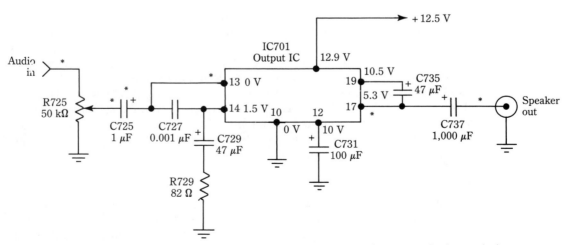

13-25 Check for distortion at X marks, then check the voltage at each pin terminal.

The resistor might be burned open or increased in resistance. These resistors should be checked if the transistors are tested or removed from the circuit. Do not replace a leaky or open power-output transistor without first checking the bias resistors.

Intermittent hum

Hum pick up might occur in the front end of preamp or output circuits. Often, a broken base resistor from ground might pick up hum. Hum can be signal traced through the entire audio circuits. If you can hear the hum, start checking for poor grounds or shielding. Intermittent hum might be caused by poorly soldered grounds or sockets between amp boards.

Hum might occur if the cassette chassis is moved. Tighten and solder all shields, chassis, and motor brackets. To check, use a small lead with alligator clips at each end and ground all shields to the metal chassis. Poor cable or screw connections might cause hum problems. A broken lead from the tuner to the amplifier might pick up hum. Check all shielded cable for proper grounds. Tighten all ground screws.

Unusual sound problems

If the volume control is rotated and only operates at one point, suspect that a control is defective or that the components that are tied to it are defective. Clean the control with cleaning fluid. If you only hear the volume at the center of the control, check for a grounded center terminal. Measure the entire resistance of the control.

If the control seems normal, check for dc voltage on the volume control. No voltage should be found on R145 or R147 (FIG. 13-26). Suspect that either C215 or

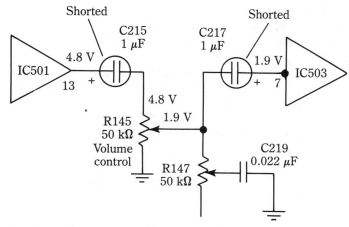

13-26 Unusual audio problems can occur if two identical capacitors are shorted on each side of the volume control.

C217 is leaking if you find voltage at the volume control. If C215 and C217 become shorted at the same time, dc voltage will appear at the top of the volume control and at the center arm. Because both of these 1-μF capacitors have the same value, it's possible that both might be leaky at the same time. The unusual problem results in the signal only being audible at one point on the volume control.

Defective speaker terminals

Intermittent or dead audio channels might result from an improperly twisted speaker cable that shorts out both speaker terminals at the rear. If inside terminals are broken, or if solder is lopped over, the problem might short out the speaker terminal to the metal chassis. Only strip back enough speaker-wire insulation and tightly twist it. Tin or solder each speaker terminal before sliding it in or wrapping it around the speaker post (FIG. 13-27).

13-27 Check the speaker terminals for loose, frayed, or twisted wires if the speaker volume is dead or intermittent.

External and old speaker wire should be replaced if the insulation will not stay on the wires. Make sure that the positive terminal is at the positive side of the speaker and that the negative is at the negative post at the rear of the cassette deck for correct speaker phase. If the speakers are not marked correctly, polarity can be made at the speakers.

Add two alligator clips to a 1.5-V flashlight cell and clip the negative terminal to one side of the speaker voice-coil terminal. Hold your left hand lightly against the speaker cone and if the positive terminal of the battery is touched to other terminal, notice if the cone pulls in or pushes out. Reverse the battery leads until the positive terminal pushes the cone outward. Mark this terminal positive, in reference to the positive terminal of the battery (FIG. 13-28). Likewise, polarize both speakers with the same method. Do not leave the positive clip on the battery. Now, connect the positive lead on the speaker to the positive terminal at the rear of the cassette deck.

Defective muting switch

In many of the larger audio amps, a muting switch might be defective and make the audio channel dead. If this is the case, no sound or line output is found in this

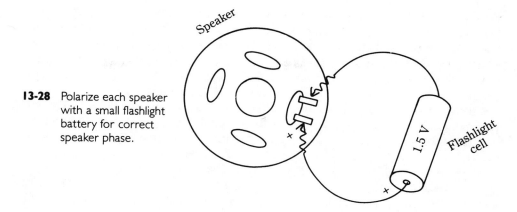

13-28 Polarize each speaker with a small flashlight battery for correct speaker phase.

channel. Often, transistor-biased circuits operate as a muting switch. Signal trace the audio to the muting transistor. Here, the output audio might be traced from pin 7 of IC500 to R521 (FIG. 13-29).

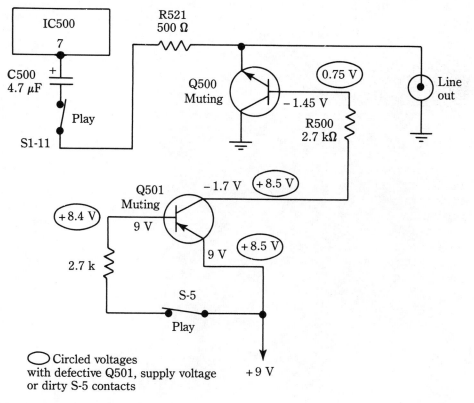

13-29 Defective muting circuits can be caused by a dirty switch, an open Q501, or an improper supply voltage at Q501.

If you find volume on one side of R521 and not on the emitter of Q500 and line output, suspect that the muting system is defective. Test Q500 and Q501 for leaky conditions. If the transistors test normal in the circuit, take voltage measurements. Transistor Q501 provides muting bias so that Q500 will stay open. If bias is removed from the base of Q500, it acts like a switch from the emitter to collector ground, and shuts off any sound at the line or speaker output.

If the bias voltage at the base terminal of Q500 becomes positive, Q500 acts as a switch. With a negative voltage at the base terminal, the switch is open and you can hear sound. Suspect a defective muting transistor or open components on the transistor terminals if this is the case. If a wire is broken off of S5 or if the contacts become dirty, Q501 will provide a positive bias voltage to the base of Q500. Make sure that all voltages are normal on both transistors. A defective switch, dirty switch terminals, or improper voltages at the collector of Q501 will cause Q500 to act as a switch and mute the output sound.

Speaker enclosures

The compact or stacked cassette deck might have large shelf or stand-by speaker enclosures. These high-powered amplifiers must be connected to high-wattage speaker systems. The 12- or 15-inch speaker is the woofer and the midrange horn-type speaker is at the top (FIG. 13-30).

13-30 Large 12- or 15-inch floor speakers can be driven by a high-powered amplifier.

Radio Shack

The speaker column might have a 3-way speaker system with a 15-inch woofer, tuned-port enclosure for deep bass response, 5-inch midrange, and 4-inch dual radial super horn tweeter. The large high-powered speaker system might consist of a 12-inch woofer, a 6-inch midrange, and a 3- or 4-inch tweeter (FIG. 13-31). These speakers might handle 100 to 160 W of power.

13-31 The large floor speaker might contain three different-sized speakers; in this case, a 12-inch woofer, a 6-inch mid-range, and a 3-inch tweeter.

Simply interchange the speaker wire terminals if one speaker column channel is dead. If the good speaker does not operate on the dead channel, the speakers are normal. If the exchanged speaker is dead, the speaker column is open. Often, the tweeter speaker will play with no woofer sound if the voice coil is blown open from excessive volume. Replace the defective speaker with one of the same physical size, magnet weight, and the same or higher wattage.

Chapter 14

Simple camcorder repairs

*Y*ou can make many different repairs to keep your camcorder taking pictures. In fact, many problems that occur in the VCR mechanism are comparable to those that are found in the camcorder. The camcorder is merely a camera and the recorder is nothing more than a smaller version of a VCR. The recorder places the video and audio on tape and plays it back for viewing with a TV and VCR.

The different formats VHS, VHS-C, and 8-mm cassettes are used in different camcorders. The VHS format was one of the first camcorders mass produced; it is very stable and produces one of the best recordings (FIG. 14-1). Although the VHS machine is quite heavy and bulky, great pictures can be taken with the VHS format. The VHS camcorder prevents accidental movement with very stable pictures. The soundtrack is at the bottom of the VHS tape.

The VHS-C format is a smaller cassette version of the VHS machine. The cassette is about $1/3$ of the size of the VHS cassette, which results in a smaller compact camcorder. The same width of tape in the VHS machine is used in the VHS-C format. Although the compact (VHS-C) cassette cannot be played directly within the VCR for viewing on the TV screen, it can be placed inside a VHS-C cassette holder for playback. The compact-disc camcorder can play through the TV antenna system.

The 8-mm format has a thinner cassette than VHS-C and it is smaller in size (FIG. 14-2). This makes the camera easier to handle and hold. In fact, today's 8-mm camcorder weighs less than 2 lbs. Although the 8-mm camcorder must be held steady for stable pictures, some manufacturers have circuits to compensate for unstable photos. The 8-mm camcorder cassette will not fit in any VHS recorder; they must only be used in 8-mm VCR recorders. The 8-mm camcorder is a camera

14-1 The VHS 10X power-zoom RCA camcorder has 2-speed record and play with pro edit features.

14-2 The RCA pro 8 (8 mm) camcorder has a CCD image sensor instead of a MOS sensor.

and a recorder, but it has no internal playback functions. The 8-mm camcorder uses a flying erase head and sound recorded within the video system.

DEFECTIVE CASSETTES

Do not try to record with broken, cracked, or damaged cassettes. You might damage, pull, or jam tape within the tape mechanism. Inspect the cassette before loading it into the camcorders (FIG. 14-3). Keep your fingers away from the tape opening so that oil from your fingers will not contaminate the tape. Inspect the back area for broken tabs. If a tab is broken out or switched out, the cassette will not record.

14-3 Inspect cassettes for possible body damage.

Keep all tape inside the cassette. If the tape is pulled out, rotate the hub until all tape is inside the plastic area. Do not throw tapes, but stack them on edges within the plastic case. Keep cassettes away from magnetic components, such as motors, TVs, and speakers. Try to keep cassettes dry at all times. Bring the cassette out of the cold and let it sit for at least 20 minutes before inserting it into the camcorder. Some camcorders have dew sensors that prevent playback or recording with moisture present. Avoid leaving cassettes in the car, in strong sunlight, or where temperatures exceed 140° F.

SNOWY/NOISY PICTURES

Clean the tape heads if the picture becomes fuzzy, snowy, streaky, or noisy. It's better to clean the tape head than to wait for major repairs. Excessive packed oxide on the tape head drum might cause pulled tape, which could damage the tape and internal parts. The heads can be cleaned with a cleaning cassette, cleaning spray, or with cleaning fluid and a chamois (FIG. 14-4).

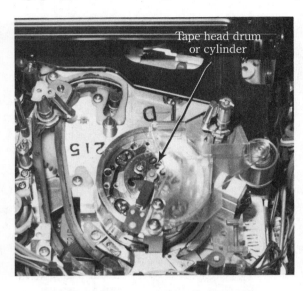

14-4 A dirty tape-head cylinder might produce snow, noise, streaking lines, and poorly colored pictures.

Some video cleaning cassettes might leave a residue and not thoroughly clean the heads. Select a cleaning cassette that will also clean spindles and rollers. Of course, it is best to clean the drum-head assembly with a chamois, soaked in cleaning fluid. Sometimes the heads can be cleaned through the cassette door, but it's best to remove the door cover.

Many door covers can be removed to get at the tape-head mechanism (FIG. 14-5). In this RCA CPR300 model, remove the two rubber fill pieces in the front cover. Two metal screws are located under the rubber pieces. Now, remove these screws and the door will slip off.

Look for the large shiny rotating drum. The tape heads are located at about the center of the drum area. The video heads might be cleaned with a can of cleaning fluid. A small plastic tube lets you spray the head area from the cassette-holder area. If you are using this method, be careful not to spray fluid all over the inside of the VTR area. Cleaning with the video head cleaning spray can is only a temporary affair.

The camcorder heads should be cleaned twice a year if it is used extensively. Remove the front cover and do a good job. The best method is to use a sponge swab or a chamois with cleaning fluid. Do not use ordinary audio cleaning sticks. The cotton material will pull and cling to the small head opening. An alcohol-dampened lint-free cloth is good. VCR tape-head cleaning kits are available at most camera and camcorder stores.

Coat the cleaning chamois stick with cleaning fluid and touch the stick to the head cylinder. Rotate the cylinder drum to the right and left while holding the cleaning stick vertically. Move the stick horizontally and not vertically, to prevent

White rubber insert

Black rubber insert

14-5 Remove the front cover to get at the heads in the RCA CPR300 by taking out the rubber wedges, then the metal screws.

damage to the head-tip assembly (FIG. 14-6). Keep your fingers off of the tape-head surface. You can damage the small head assembly if you pull it up and down. Just rotate the head horizontally while holding the coated stick stationary. Clean oxide dust from all heads and the metal drum surface.

While you're at it, clean the various guides, rollers, and loading platforms with cleaning fluid. Wipe up all signs of oxide dust. Visually inspect the tape-drive

14-6 When you clean the tape-head assembly of a JVC VHS-C camcorder, move the chamois stick horizontally, not vertically.

Tape head drum

areas for brown oxide dust and dirt. In addition to a poor picture and sound, improper clean up of the pressure rollers and the capstan causes tapes to be eaten from the cassette (FIG. 14-7). Wipe off the capstan with a damp cloth.

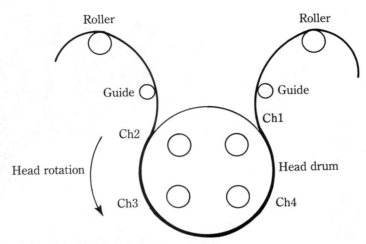

14-7   Clean tape guides, rollers, and spindles in the tape path when you clean the tape heads.

ACCIDENTAL ERASE

You might accidentally insert a cassette in the camcorder and erase good pictures while taking other shots. To keep cassettes from being ruined or used, knock out the small plastic tab in the rear of the cassette. In cassettes that you want to keep, knock out the small tab.

The erase tab is located in the VHS, Beta, and VHS-C cassettes. In the 8-mm cassette to protect recording from erasure, slide the red safety switch in the direction of the arrow. To record again, reset the switch (FIG. 14-8). If you want to record over a VHS, VHS-C, or Beta cassette that has the plastic tab removed, place a piece of tape across the opening.

NO TAPE ROTATION

After loading the camcorder with a cassette and the camcorder will not record or play, suspect that no power is applied to camcorder. Make sure that the battery is installed properly. Do any of the indicator lights come on? Some camcorders have battery indicators that show if the battery is getting weak. Insert an ac adapter instead of a battery to see if the camcorder will operate. Make sure that the power switch is on. If the green light is flashing inside view finder, the battery needs to be charged.

Check for a blown fuse if nothing moves or lights. If the unit operates on ac and not batteries, charge the battery. It's nice to have two batteries, one in the camcorder and the other charged up. The power LED should be on if the power

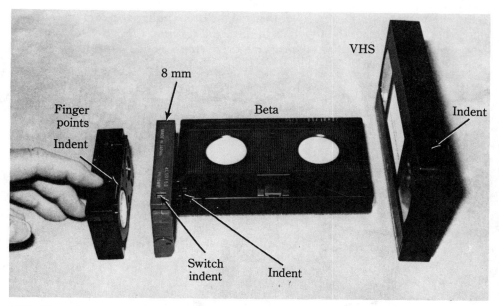

14-8 To avoid accidentally erasing a favorite cassette, remove the tabs or slide the red safety switch.

switch is on. After replacing the ac adapter and fuse, suspect that a power switch or motor is defective. Check the power switch and circuits with the DMM.

BROKEN ON/OFF SWITCH

After checking for a blown fuse and checking the voltage applied to the camcorder, inspect the on/off switch. Although these switches do not cause too many problems, the camcorder might have been dropped on the switch area. Many camcorder problems result from how the unit was handled. VHS camcorders are quite heavy and should be set down; do not hand the camcorder to another person. Replace switch with the original if it is defective. Besides broken switches, inspect the electronic viewfinder for breakage. If the glass rattles around when the camcorder is handled, the viewfinder tube might be broken. Before thinking about replacing it, check for an estimate. These tubes are rather expensive.

PULLS TAPE

The dirty tape path, rollers, spindles, capstan, and tape-head drum might cause the tape to pull out and wind around moving parts. Oil on moving parts, capstan, and reels might cause the tape to pull. Sticky substances on the drum or sound heads might eat the tape. Do not overlook the possibility that a cassette might be defective. First, clean the entire tape path with alcohol and chamois or cloth. Then, try another cassette.

If the tape is pulled from the cassette and tangled in the mechanism, try to remove the cassette. Shut off the camcorder at once. If the cassette cannot be removed, remove the bottom cover to get at the drum pulley and capstan flywheel (FIG. 14-9). Try to rotate the flywheel in reverse if it doesn't move. Sometimes the tape will wrap tightly around the capstan drive area.

14-9   Carefully inspect the pulled out tape and rewind it by rotating cassette hub assembly.

Be careful when removing excess tape not to damage the video heads on the drum assembly. Sometimes the tape must be cut and removed in pieces. Of course, the cassette is damaged unless it can be spliced. If the tape has several hand wrinkles or torn areas, it's best to discard it. After the cassette is removed, make sure that all excess tape is removed from the various moving parts.

Sometimes the excess tape can be fed back into the cassette if too much tape is not released. Press the stop button and power is applied. The tape might feed back into the cassette as the rewind period occurs. If the tape has jammed the cassette and machine, the tape won't move. Make that sure all tape is wound back into the cassette before attempting to use it. Both front and back covers might have to be removed to get at the flywheel, pulleys, and cassette.

JAMMED CASSETTE

The defective or broken cassette case might cause the tape to jam or not wind or rewind. Closely inspect each cassette that is inserted into the camcorder. Sometimes the damaged cassette will not load. A cracked or broken cassette should be discarded. Not only will it cause improper tape movement, but it might jam the camcorder mechanism. A cassette wound too tight might jam or pull the tape. If in doubt, always try another cassette.

CASSETTE WILL NOT LOAD

In some camcorders, the cassette is placed in the cassette holder and the door is manually shut. In other camcorders, the cassette is placed in holder and loaded electronically with a loading motor. Inspect the loading area for foreign material or a possible defective cassette if the cassette will not seat properly. Make sure that the loading platform is not bent. Check for a broken or damaged door.

Listen for the operation of the loading motor in camcorders that pull the cassette in and close the door automatically. If the motor is not rotating, suspect that a loading switch lever is broken or bent. The front and rear covers might have to be removed to get at the loading motor. Check the loading motor for a jammed gear assembly, or a loose or broken drive belt. Sometimes a gum wrapper or cigarette paper might get inside, wrap around the loading belt, and prevent cassette loading.

Check the continuity of loading motor winding with RX1 ohmmeter range (FIG. 14-10). No reading might indicate that the motor armature winding or brushes are open. Measure the voltage applied to the loading motor terminals (1.5 to 5 Vdc). Inspect the motor plug and wire for poor contacts or breakage. A defective motor-drive IC or transistor might not apply voltage to the loading motor. Measure the output voltage to the motor and the supply voltage to the motor IC or transistor.

14-10 Check the continuity of the loading motor with the RX1 range of the DMM.

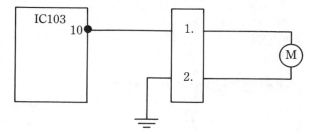

STICKY BUTTONS

Most of the top-operation control buttons are under a plastic sleeve and cause little problem. Zoom and side buttons might become damaged or the plastic against the plastic might cause the buttons to stick. Lightly spray cleaning fluid around the outside buttons. Work the switch back and forth to ease the sticky button. If the buttons are damaged by something striking the switch areas or if the camcorder dropped, they might have to be replaced. Most surface-mounted buttons can be cleaned with cleaning fluid.

ERRATIC OPERATION

If a zoom button is pressed and the lens stops moving, suspect a dirty switch button or control. Remove the front plastic cover to get at these controls (FIG. 14-11). Inspect the switches for broken plastic. If the switches appear to be normal, clean the switches with cleaning fluid.

14-11 Clean switches and controls that seem to be intermittent.

Place the small plastic nozzle into the switch area. Release some fluid. Work the switch back and forth to clean the switch contacts. If the switch will not clean up, inspect the terminals for poor contact wires. Notice if the terminal connections are poorly soldered. Retouch the switch contacts with the soldering iron. The switch assembly might have to be removed or pulled back to get at the switch contacts. If the switch will not clean up, replace it with the original part number.

ERRATIC AND SLOW SPEED

Erratic speed might be caused by a loose, oily, or cracked drive belt. Inspect the belt for worn or cracked areas. Older belts have a tendency to crack and stretch. Clean the belt with alcohol and a rag if it looks okay. Actually remove the belt from the drive pulley and flywheel to do a good clean up. If the belt is checked or loose, replace it (FIG. 14-12).

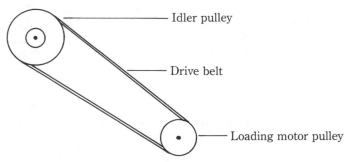

14-12 The cracked or loose drive belt should be replaced to prevent erratic or slow speeds.

Check the belt for shiny or dull surfaces. A belt that is slipping will show a smooth side on a dull surface. The dull surface should be against the pulley and the motor drive pulley. If the belt shows signs of slippage (shiny grey areas), replace it.

Before replacing the defective belt, clean the pulleys, idlers, and capstan-drive areas with alcohol and a cleaning stick. Notice if small particles of rubber are sticking to the flywheel area. Keep oil off of the belts and fingers. Clean the belt area with alcohol and a cloth, after installing the new belt, to remove oily finger marks. Make sure that bearings and surfaces are clean after lubrication. Changing the belt when cleaning might solve many speed problems. Erratic speed problems might be caused by a defective brake-release belt (FIG. 14-13).

14-13 Erratic speed problems might result from a nonreleasing brake assembly.

Weak audio or no audio might be caused by dirty audio tape heads. VHS and VHS-C camcorders (FIG. 14-14) have separate audio heads. Clean the audio tape head with alcohol or a cleaning stick. If you have a regular tape-head cleaning kit, use the same cleaning fluid. Sometimes, the oxide might be packed into the gap area, which keeps the head from recording or playing.

Notice if the video is normal within the viewfinder and if the cassette is played in the VCR. Check the sound output at the earphone jack. Some camcorders have this feature and others do not. Most camcorders have one large audio IC. Take a critical voltage measurement at the sound IC. Do not try to adjust audio azimuth adjustment without a test tape and scope. Leave this adjustment up to the professional electronic technician. Remember, the audio within the 8-mm camcorder is recorded in the video section of the tape and these camcorders do not have a separate audio tape head.

NO RECORD

Does the camcorder play a cassette without any problems? Can you see a picture in the viewfinder? If the record button will not seat, suspect that the tab is pulled

14-14 Weak or no audio recording might result from a packed audio tape head. Clean it with alcohol and a cleaning stick.

from the cassette. Try another cassette. Switch over the red lever within the 8-mm cassette.

Make sure that the cables are correct if you can see the normal picture in the viewfinder and not on the TV screen. Are the cables plugged into the correct jacks when the camcorder is played directly through the antenna system of the TV? Recheck the recording with the VCR unit. Doublecheck the audio/video output cables of the camcorder to the audio/video-in jacks of the VCR.

Check to see if the camcorder is actually in the record mode. Make sure that you press the right buttons. Clean the tape heads if the unit is not recording or playing. Critical recording-circuit repairs should be made by the electronic technician.

NO FOCUS CONTROL

The camcorder can focus manually or electronically. The lens assembly will rotate automatically if the camcorder is self-focusing. Suspect jammed gears, a bad focus motor, or bad circuits if the focus-lens assembly will not move. Inspect the motor gear train for foreign material. Check for broken or stripped gears. Measure the voltage that is applied to the focus motor (FIG. 14-15). Check the motor winding continuity with the ohmmeter on the RX1 setting. Notice if the focus-sensor assembly, below the lens assembly, is covered or damaged.

NO POWER VOLTAGE

Try operating the camcorder on batteries or on ac power. If the unit works on batteries and not on ac, suspect that a low-voltage power supply is defective. Check for open fuse with no power from either source. If the unit works on ac power,

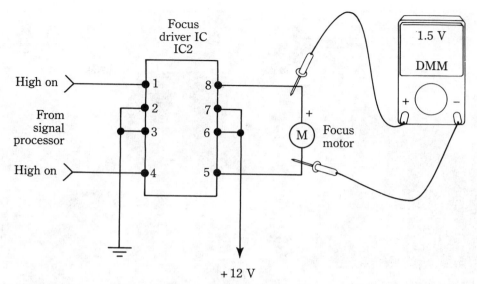

14-15 If the camera is experiencing problems with the focus control, measure the voltage applied to the focus motor.

but not on batteries, check the contacts on the external battery jacks. An open fuse in the power supply might prevent voltage from reaching the camcorder (FIG. 14-16).

Inspect the voltage-battery clips where the battery or ac power jack snaps into position for bad or dirty connections. Measure the voltage at the ac power supply (FIG. 14-17). Some units have interlocks that will not show a normal reading

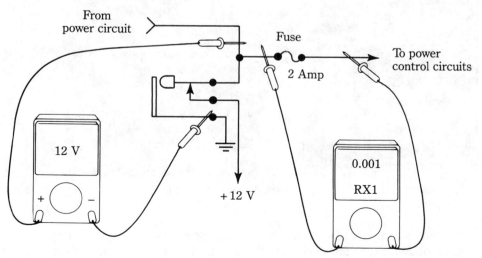

14-16 Check for an open fuse in the power supply or power circuits if the camera has no power voltage.

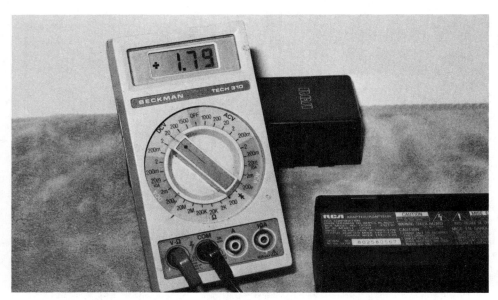

14-17   Measure the voltage at the battery or ac power-supply adaptor.

until they are snapped into place. Remove the bottom cover of power supply. Check for an open fuse (FIG. 14-18). Test all transistor regulators and diodes with the diode test of the DMM.

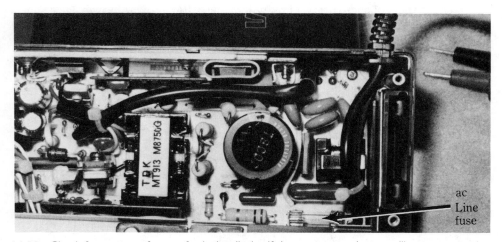

ac
Line
fuse

14-18   Check for an open fuse or for leaky diodes if the ac power adaptor will not operate the camcorder.

Check the power circuits inside the camcorder if neither the battery nor the ac pack operates the camcorder. Set the unit to the camera mode and check for a picture in the viewfinder. Do any indicators light? If not, go to the power-distribution circuits inside the camcorder. Check for open fuse. Measure the 9- and 12-V source. If you find no voltages at all, check the transistor power switches (FIG. 14-19). Test all transistors or IC regulator components with the DMM. Let the professional technician replace any control ICs if they are defective .

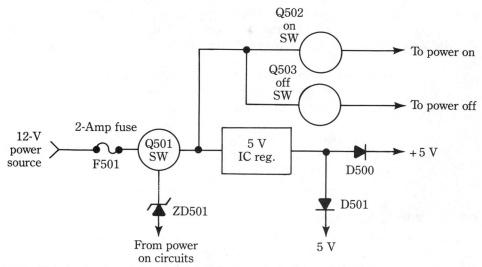

14-19   Test the transistors and measure the voltage on the IC regulators in the ac power adaptor.

BATTERY WILL NOT CHARGE

The ac battery charger can be checked by placing a 10-Ω load across the charging terminals. This will switch on the battery adapter, charger interlocks, and the output voltage (FIG. 14-20). Measure the output voltage across the 10-Ω resistor. Suspect that a battery is defective if the charging voltage is present. In most adapter battery-charging units, the battery clips into position. Keep the batteries charged at all times. A defective battery will not hold a charge.

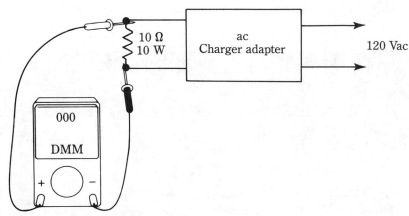

14-20   Check the ac charger/adaptor voltage by placing a 10-Ω load across the terminals.

Notice if the small charging LED is on. If no light and voltage is at the output terminals, remove the cover of the charger and check the fuse. Check continuity of the transformer windings and the silicon rectifiers. Inspect the charging silicon diodes. If the charging LED is on, the unit should be charging because this circuit is in the output-charging circuits. Clean all battery-charger contacts with alcohol and a cloth. The various camcorder operating voltages are listed in TABLE 14-1. Make sure that the battery charger is functioning before you replace the dead battery.

Table 14-1. Operation Voltage of Various Camcorders

Name	Model	Voltage	Format
Cannon	CA-E2A	6 V	8 mm
JVC	GRC-7U	9.6 V	VHS-C
RCA	CPR300	12 V	VHS
Sony	CCD-M8E	6.3 V	8 mm

BATTERY REPLACEMENT

If the batteries will not charge after you connect them to the charger for several days, suspect that the battery is dead. Make sure that the battery terminals are clean. Camcorder batteries can now be obtained from the manufacture parts depot, camcorder/camera stores, local electronic wholesale houses, and mail-order stores. Write down the old battery part number. You might be able to replace the battery with a universal battery replacement. If possible, obtain an original battery from the camcorder manufacturer or parts supply.

DEFECTIVE MOTORS

Basically, three motors (drum, capstan, and zoom motors) are found in most camcorders. Larger camcorders might contain drum, capstan, loading, auto-focus, iris, and zoom motors. The auto-focus, iris, zoom, and loading motors might not be found in the small camcorders. Some 8-mm camcorders have only drum, capstan, and loading motors (FIG. 14-21).

14-21 Locate the suspected motor on the camcorder chassis.

The motor can be checked with ohmmeter tests, voltage measurements, and with an external voltage applied to the motor terminals. Check motor continuity with the RX1 range of the DMM (FIG. 14-22). Measure the voltage across the motor terminals. The dc voltage might be applied from the IC or transistor driver circuits (FIG. 14-23). The drum-cylinder and capstan motors might have complicated servo and drive circuits, but the loading and focus motors are simple dc motors.

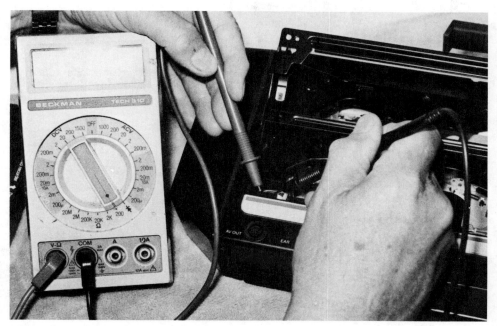

14-22   Check the motor with the RX1 ohmmeter range of the DMM.

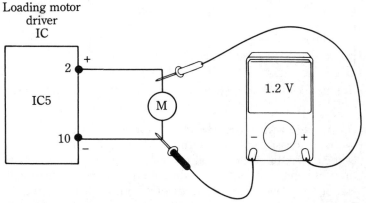

14-23   Check for low dc voltages across the motor terminals to determine if the motor is defective.

No drum movement

The cylinder or drum motor is quite complicated and should only be serviced by a professional. You can check the drum-motor windings for continuity and voltage applied to the drum servo IC. The drum motor is located under the top drum (FIG. 14-24).

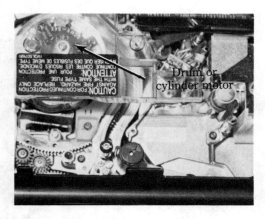

14-24   The drum or cylinder motor is located under the tape-head (top) assembly.

CASSETTE WILL NOT LOAD

The loading motor might eject, load, and unload the video cassette. Just press the eject button on most camcorders and the loading door will open to eject or resume the cassette. After inserting the cassette, the loading motor might close the door. The loading motor is at the side and is controlled by the system-control IC, motor-drive IC, or by early camcorder transistors (FIG. 14-25).

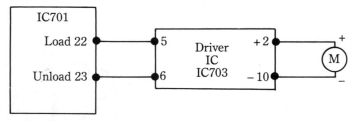

14-25   The loading motor operates from a driver IC and system-control processor.

Check the voltage across the loading motor when you press the button. If you find no voltage across the motor terminals, check the driver output and supply voltage at the driver IC. If no signal voltage is at pins 5 and 6, IC701 might be defective. Often, if other functions are controlled and operate normally from the control IC, either the driver, IC, motor, or eject button is defective. Voltage and continuity tests on the motor terminal might determine if motor is defective.

Zoom motor hums/no movement

The zoom motor brings the image close or far away from the lens. The power zoom (PZ) or zoom motor is mounted on the lens assembly. The zoom motor is geared to the lens assembly. If the motor hums with no movement, suspect that foreign material is in the gear train or that the gear train is broken and damaged. The zoom motor and lens assembly is out in the open where it can easily be damaged if it is slammed against trees, buildings, or poles while shooting different scenes.

The zoom motor might be driven with transistors or ICs. The tele and wide-angle switches insert or place different resistance into the drive circuit to change the direction of the zoom motor (FIG. 14-26). The zoom motor is a dc motor and it can be checked with voltage and continuity tests across the motor terminals. Press either the tele or wide switch when you take the tests. Measure the supply voltage that is applied to the zoom driver IC.

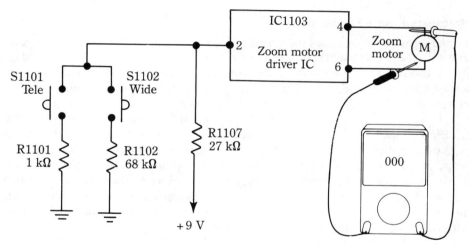

14-26 The dc zoom motor is operated from a dc voltage supplied by driver IC1103.

No auto focus

Infrared rays are provided to control the auto-focus circuits. Infrared rays are emitted from the infrared LED to the image or object and reflected back to the camcorder. Two photodiodes are used in this system to detect the infrared rays reflected back from the object, and convert to the current signals then to a control voltage. Check the infrared LEDs under the lens assembly with the infrared detector from chapter 1.

The auto-focus motor is controlled from an auto-focus processor and driver IC or transistors (FIG. 14-27). If the infrared LED system is working, check the voltage across the auto-focus motor terminals. If the voltage across the motor terminals is okay and the motor isn't moving, suspect that a motor winding is open or

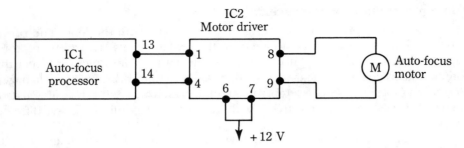

14-27 The auto-focus motor is voltage controlled with driver IC2 and IC1.

that the motor connections are poor. Check the motor continuity with the RX1 range of the DMM. If there is no motor voltage, check the output at terminals 8 and 9 and check the supply voltage at pins 6 and 7. Replace the driver IC if you find supply voltage and no motor-drive voltage.

No capstan motor movement

The capstan motor operates the various mechanical assemblies with belt or geared movements. The capstan motor operates in play, record, rewind, fast-forward, and search modes. The motor is operated from a dc voltage that is supplied from drive ICs or transistors and the system-control IC. Sometimes the drum and capstan motors are fed from the same source.

The capstan motor must be able to operate in forward and reverse directions, at different speeds. If the speed is erratic or slow, suspect that the belts are slipping or that they are cracked or worn. if they are still slow after you replace the belts, check the motor circuits. The voltage applied in play might be around 2 Vdc; in fast forward or rewind, it might be 4.6 to 5 Vdc. Check the voltage at the motor terminals to determine if the motor or driver IC is defective. Measure the motor continuity at the terminals (FIG. 14-28). All small motors must be replaced with original part numbers if they are defective.

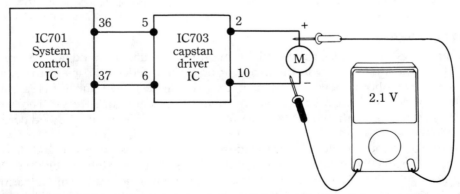

14-28 Check voltages on the capstan motor terminals to determine if the motor is open or defective.

NO PLAYBACK THROUGH TV ANTENNA

Erratic or no playback of the camcorder through the TV set might be caused by poor cable connections. Doublecheck all male connectors. Sometimes the center-shielded wire will break. Remove each cable and inspect the shield and center-wire connections. Check the channel setting of the matching box and of the RF adapter (FIG. 14-29). Sometimes the shield might not make good contact with outside metal plug. Flex the cable to check for possible breaks or poor connections.

14-29 Check and tighten all connections on the cable and RF adaptor if there is no playback through TV antenna.

NO ERASE

If the picture and audio seem to be a jumbled mess, check for erase problems. Clean the erase heads on VHS and VHS-C camcorders (FIG. 14-30). Check for the voltage on the flying erase head in 8-mm camcorders. Measure the continuity of FE head with the ohmmeter. Clean the drum-head assembly. Monitor the bias voltage with an oscilloscope.

REMOVING COVERS

Be careful if removing covers to get at the defective components. Place small screws in a saucer so that you don't lose them. Look for different lengths of screws and note what side they come out of. Write down or draw a rough picture of where the different screws and bolts are located. Place small components in a row so that they can be replaced.

The front door of most camcorders can be replaced with one or two screws and you can slip off the plastic cover to clean the heads and inspect the VCR components (FIG. 14-31). Two screws might hold the top, and the bottom edge would then slide out. Often, the back and front covers are separate and can be removed

14-30 Jumbled recordings might occur with dirty erase heads in the camcorder.

14-31 Remove the door cover to get at the heads.

individually. Carefully inspect all areas for tiny screws. Use small Phillips and jewelers' screwdrivers to remove tiny screws. Place the front-cover components at one side and the back-cover components at the other side.

A simple troubleshooting chart is found in TABLE 14-2. Check TABLE 14-3 for additional camcorder problems.

Table 14-2. Preliminary Trouble Checklists

Problem	Check
No power	Battery pack.
	ac adapter correctly installed.
	Battery charged.
	Substitute another battery.
	Power switch on.
	Rec-safety switch.
Cassette Loaded-No power	Cassette tab missing.
	Rec-safety switch defective.
	Power LED on.
	Listen for loading motor.
Will not record	Cassette tab missing.
	8 mm cassette switched at back.
	Push Rec/Standby switch.
While recording camcorder unloads, shuts off	At end of tape.
	Defective tape-end switch.
In recording color	Inspect white balance setting.
Different than actual color	
External microphone does not record	Inspect microphone switch.
	Check mic cable and plug.
No playback while tape is running	Set TV/video switch to *video*.
Playback picture blurred and noisy	Dirty video head.
	Worn head.
	Defective head.
Noise bars in picture	Reset tracking control.
	Dirty tape head.
	Defective cassette.
Tape stops during fast forward and rewind	Display set to memory.
	Time set to memory.
Will not fast forward or rewind	Tape at end.
	Bad drive belt.
	Slipping drive belt.
Dew light flashing	Moisture in camcorder-let set one hour.
Only one or two indicators on	Low battery.
	Charge battery.
	Substitute another battery.
Green light flashing in view finder	Battery low.
	Recharge or replace.
Cassette starts to load . . . but immediately shuts down	Camcorder at end of tape.
	Tape not engaged.
	Outside infrared source-triggering unit.

Table 14-3. Camcorder Troubleshooting Chart

Trouble	Check	Remedy
No power to camcorder	Set switch to On and to camera—EVF raster	No EVF raster—check EVF circuits.
	Have EVF raster; no 9- or 12-V source	Check transistor and IC regulators in power lines.
	Has 9 and 12 V—no 5-V source	Check 5-V regulator transistor or IC.
No capstan motor operation	Check 9- or 12-V source	Repair power source.
	No fast-forward/rewind operation	Check processor and IC.
	No motor rotation	Check voltage at motor terminals.
		Check continuity of motor.
		Check supply voltage to driver IC or transistor.
		Replace defective driver IC or motor.
Cassette will not eject	Power on—no voltage	Check ac power supply replace battery.
	When press eject switches is voltage present to loading motor?	Check power on source.
		Check voltage at loading motor IC and motor terminals.
	Loading motor operating sound.	Check IC ejection system. Check voltage at loading motor.
		Check eject mechanism if motor operating.
No loading motor operation	Cannot hear motor running.	Check voltage across motor terminals.
	No voltage at motor terminals.	Measure voltage at output terminals and input terminals.
		Check supply voltage to driver IC.
		Check eject switch and clean up terminals.
Abnormal drum motor rotation.	Drum servo speed-control circuit.	Check voltage on system control IC to drum motor.
		Check voltage source to control IC.
		Check for drum PG and FG waveforms.
		Replace drum motor assembly.
No dew sensor operation.	Inspect dew sensor or sensor IC.	Check voltage across dew sensor.
		Replace dew sensor if voltage present.
	No dew sensor voltage.	Check IC sensor voltage to dew sensor.
		Check supply voltage to dew sensor IC.
		Replace dew sensor IC.
No video recording.	Look in EVF for picture.	EVF lights, but no video.
	Set select switch to camera and record.	Only snow or white raster.
	Inspect tape heads.	Clean tape heads.
Snowy, noisy lines and poor color.	Tape heads	Clean tape heads and guides.
		Inspect drum for damage.
No picture.	Set select SW to camera	Only random noise.
	No noise seen in EVF.	Check luma IC and sync circuits.

	Noise seen but no picture.	Check voltage source to green, cyan, white and yellow IC circuits.
		Check luma circuits.
Noisy playback.	Check video heads clogged.	Clean video heads and guide assemblies.
	Recording amplifier	Check voltage to recording amplifier.
		Check record/camera switching.
	Control head	Defective or clogged control head.
		Check control recording circuit.
No EVF raster.	Electronic view finder does not light up (EVF).	Check supply voltage to EVF circuits.
		Check EVF tube socket.
		Measure HV to EVF tube anode.
		Check horizontal drive circuits.
		Check horizontal sync.
	No vertical sweep—white line.	Check voltage of vertical deflection IC or transistors.
		Check vertical yoke winding.
		Replace defective vertical output sweep IC.
	No video—white screen on EVF.	Check video amp IC.
		Measure voltage on video amp IC.
		Check voltage on video-amp transistors.
		Replace video amp transistors.
	EVF CRT heaters do not light up.	Check CRT socket.
		Remove socket and check contents of tube heaters.
		Check for poor winding of flyback transformer—or open connections.
No ac adaptor operation.	Connect 10 Ω 10-W resistor across output.	Measure voltage in adaptor.
	Low or no voltage.	Check fuse. Test IC, transistor regulator, and diodes.
		Inspect ac cord.
		Check power-transformer winding continuity.
		Clean all switches.
No ac adaptor charger.	Check output voltage with resistor load.	If output voltage normal suspect charging circuits.
	IC latch and timer IC.	Check voltage on latch and timer IC.
	No charge LED.	If LED does not light, check charging circuits.
	Charges no LED light.	Check zener diodes, transistor and LED charge indicator circuits.
Battery does not operate long.	Defective battery	Discharge battery completely and charge again.
		Check battery voltage to see if fully charged.
		If won't hold charge, replace battery.

Appendix

Manufacturers of consumer electronics equipment

AUDIO CASSETTES

AFCO ELECTRONICS INC.
471 Roland Way
Oakland, CA 94621

AIWA
800 Corporate Dr.
Mahwah, NJ 07430

ALARON INC.
185 Park St.
P.O. Box 550
Troy, MI 48099

ALPINE ELECTRONICS OF AMERICA
19145 Gramercy Pl.
Torrance, CA 90501

AMERICAN ACOUSTICS, INC.
12841 Western Ave.
Garden Grove, CA 92641

AMERICAN AUDIO CORP.
636 Forbes Blvd.
S. San Francisco, CA 94080

ARA MANUFACTURING CO.
606 Fountain Pkwy.
P.O. Box 534002
Grand Prairie, TX 75053

ARTHUR FULMER ELECTRONIC DIV.
122 Gagoso
Memphis, TN 38101

AUDIOVOX CORP.
150 Marcus Blvd.
Hauppauge, NY 11788

BLAUPUNKT
2800 S. 25th Ave.
Broadview, IL 60153

CLARION CORP. OF AMERICA
5500 Rosecrans Ave.
Lawndale, CA 90260

CONCORD SYSTEMS, INC.
25 Hale St.
New Buryport, MA 01950

CRAIG CORP.
14450 Industry Circle
La Mirada, CA 90638

DAEWOO ELECTRONICS
100 Daewoo Pl.
Carlstadt, NJ 07041

DENNON AMERICA
222 New Rd.
Parsippany, NJ 07054

ELECTRONIC INDUSTRIES
16940 Vincennes
S. Holland, IL 60473

FORTUNE STAR PRODUCTS
12 W. 23rd St.
New York, NY 10010

FUJITSU CORP. OF AMERICA
19281 Pacific Gateway Dr.
Torrance, CA 90502

GRUNDIG/GR ELECTRONICS
Glenpointe Center E
Teaneck, NJ 07686

HANAHASHEYA LTD.
39 W. 28th St.
New York, NY 10001

J-CORP CORP.
1031 Northern Blvd.
Baldwin Harbor, NY 11510

JENSEN INC.
4136 N. United Pkwy.
Scheller Park, FL 60176

JVC AMERICA
41 States Dr.
Elmwood Park, NJ 07407

K & K MERCHANDISING GROUP
10-27 45th Ave.
Long Island, NY 11101

KENDALE TECHNOLOGY CORP.
4185 NW 77th Ave.
Miami, FL 33176

KENWOOD ELECTRONICS
2201 East Dominguez St.
Long Beach, CA 90810

KRACO ENTERPRISES, INC.
505 E. Euclid Ave.
Compton, CA 90224

MARANTZ CO.
20525 Nordhoff St.
Chatsworth, CA 91311

MCI
23 NW 8th Ave.
Hollandale, FL 33009

MIDLAND INTERNATIONAL CORP.
1690 N. Topping
Kansas City, MO 64120

MITSUBISHI
5757 Plaza Dr.
P.O. Box 6007
Cypress, CA 90630

NAKAMICHI USA, CORP.
19701 S. Vermont Ave.
Torrance, CA 90502

PANASONIC AUTO DIV.
One Panasonic Way
Secaucus, NJ 07094

PHILLIPS AUTO RADIO DIV.
230 Duffy Ave.
Hicksville, NY 11802

PIONEER ELECTRONICS USA
5000 Airport Plaza Dr., Box 1540
Long Beach, CA 90801

SANSUI CORP.
1250 Valley Brook Ave.
Lyndhurst, NJ 07071

SANYO ELECTRIC INC.
1200 W. Artesen Blvd.
Compton, CA 90220

SONY CORP. OF AMERICA
Sony Park
Park Ridge, NJ 07074

SPARKOMATIC CORP.
Rts. 6 and 209
Milford, PA 18337

TANCREDI DIV. KVKJE PACIFIC
2318 E. Del Amo Blvd.
Compton, CA 90220

TZL INTERNATIONAL CORP.
1523 NW 79th Ave.
Miami, FL 33126

ULTIMATE SOUND
19330 E. San Jose Ave.
City of Industry, CA 91748

VECTOR RESEARCH
20608 Nordhoff St.
Chatsworth, CA 91311

YAMAHA ELECTRONICS CORP.
6660 Orangethorpe Ave.
Buena Park, CA 90620

CAMCORDERS

AIWA
35 Oxford Dr.
Moonachie, NJ 07074

CANON
One Canon Plaza
Lake Success, NY 11042

CHINON
43 Fadem Rd.
Springfield, NJ 07081

CURTIS MATHES
1220 Champion Circle
Carlton, TX 75006

ELMO
70 New Hyde Park
New Hyde Park, NY 11040

FISHER
1200 W. Walnut, Box 9038
Compton, CA 90224

GENERAL ELECTRIC
Box 1976
Indianapolis, IN 46206

GOLDSTAR
1050 Wall St.
Lyndhurst, NJ 07071

HITACHI
401 W. Artesia Blvd.
Compton, CA 90220

INSTANT REPLAY
2951 S. Bay Shore Dr.
Coconut Grove, FL 33133

J.C. PENNEY, NAT'L PARTS CENTER
6840 Barton Rd.
Morrow, GA 30260

JVC
41 Slater Dr.
Elmwood Park, NJ 07407

KODAK
343 State St.
Rochester, NY 14650

KYOCERA
411 Sette Dr.
Paramus, NJ 07652

MAGNAVOX/PHILLIPS CONSUMER ELEC.
(Philco and Sylvania) Box 967
Greenville, TN 37944-0967

MINOLTA
101 Williams Dr.
Ramsey, NJ 07446

MITSUBISHI
5757 Plaza Dr., Box 6007
Cypress, CA 90630

NEC
1255 Mechael Dr.
Woodale, IL 60191

NIKON
623 Stewart Ave.
Garden City, NY 11530

OLYMPUS
Crossways Park
Woodbury, NY 11797

PANASONIC
One Panasonic Way
Secaucus, NJ 07094

PENTAX
35 Muerness Dr., E.
Englewood, CO 80112

QUASAR
1325 Pratt Blvd.
Elk Grove Village, IL 60007

RADIO SHACK, NAT'L PARTS CENTER
900 E. Northside Dr.
Fort Worth, TX 48106

RCA, THOMSON CONSUMER ELEC.
Box 1976
Indianapolis, IN 46286

RICOH
5 Didrick Pl.
West Caldwell, NJ 07006

SANYO, SFS CORP.
1200 W. Walnut St. (Box 9038)
Compton, CA 90224

SEARS
Sears Tower
Chicago, IL 60684

SHARP
Sharp Plaza
Mahwah, NJ 07430

SONY
Sony Drive
Park Ridge, NJ 07656

TECHNIKA
353 Route 46 W.
Fairfield, NJ 07006

TOSHIBA
82 Totowa Rd.
Wayne, NJ 07470

VIVITAR
1630 Stewart St.
Santa Monica, CA 90406

ZENITH
1900 Austin Ave.
Chicago, IL 60639

CD PLAYERS

AKAI AMERICA LTD.
800 W. Artesia Blvd.
Compton, CA 90220

DENNON AMERICA, INC.
222 New Rd.
Parsippany, NJ 07054

FISHER CORP.
21314 Lassen St.
Chatsworth, CA 91311

GENERAL ELECTRIC
Portsmouth, VA 23705

HITACHI
401 W. Artesia Blvd.
Compton, CA 90220

JVC AMERICA CO.
41 Slater Dr.
Elmwood Park, NJ 07407

KENWOOD ELECTRONICS
1315 E. Watson Center Rd.
Carson, CA 91745

KYOCERA INTERNATIONAL
7 Powder Horn Dr.
Warren, NJ 07060

LUXMAN DIV.
3102 Kashiwa St.
Torrance, CA 90505

MAGNAVOX
NAP Box 6950, I40 & Straw Plains Pike
Knoxville, TN 37914

MARANTZ CO.
20525 Nordhoff St.
Chatsworth, CA 90221

NAD, INC.
675 Canton St.
Norwood, MA 02062

NAKAMICHI USA CORP.
19701 S. Vermont Ave.
Torrance, CA 90502

NEC HOME ELECTRONICS
1401 W. Ave.
Elk Grove Village, IL 60007

ONKYO USA CORP.
200 Williams Dr.
Ramsey, NJ 07446

PANASONIC CORP.
One Panasonic Way
Secaucus, NJ 07094

PIONEER ELECTRONICS
5000 Airport Plaza, Box 1540
Long Beach, CA 90801

QUASAR CO.
9401 W. Grand Ave.
Franklin Park, IL 60131

RCA CONSUMER ELECTRONICS
600 N. Sherman Dr.
Indianapolis, IN 46201

SAE
1734 Gage Rd.
Montebello, CA 90640

SANSUI ELECTRONICS
1250 Valley Brook Ave.
Lyndhurst, NJ 07071

SANYO ELECTRIC INC.
1200 W. Artesia Blvd.
Compton, CA 90220

H. H. SCOTT INC.
20 Commerce Way
Woburn, MA 01888

SHARP ELECTRONICS
10 Sharp Plaza
Paramus, NJ 07652

SONY CORP.
Sony Drive
Park Ridge, NJ 07074

STUDER REVOX OF AMERICA
1425 Elm Hill Pike
Nashville, TN 37210

TECHNICS
One Panasonic Way
Secaucus, NJ 07094

TOSHIBA AMERICA
82 Totowa Rd.
Wayne, NJ 07470

YAMAHA ELECTRONICS CORP.
6660 Orangethorpe Ave.
Buena Park, CA 90620

COMPACT SYSTEMS

ALARM INC.
Box 550
Troy, MI 48099

BANG & OLUFSEN OF AMERICA
1150 Feehanville Dr.
Int. Prospect, IL 60056

BRENTWOOD ELECTRONICS
256 W. Ivy Ave.
Inglewood, CA 90302

DENNON AMERICA INC.
222 New Rd.
Parsippany, NJ 07054

EMERSON RADIO CORP.
North Bergen, NJ 07047

FORTUNE STAR PRODUCTS
1200 23rd St.
New York, NY 10610

HANIMEX INC.
3125 Commercial Ave.
Northbrook, IL 60062

HITACHI SALES CORP.
401 W. Artesia Blvd.
Compton, CA 90220

JULIETTE ELECTRONICS
4615 NW 77th Ave.
Miami, FL 33166

LLOYDS ELECTRONICS
180 Rariton Center Pkwy.
Edison, NJ 08818

MAGNAVOX CONSUMER ELECTRONICS
Box 6950
Knoxville, TN 37914

MARANTZ CO.
20525 Nordhoff St.
Chatsworth, CA 91311

MITSUBISHI ELECTRIC SALES
3030 E. Victoria
Rancho Dominguez, CA 90221

PANASONIC
One Panasonic Way
Secaucus, NJ 07094

PHILCO CONSUMER ELECTRONICS
Box 6950
Knoxville, TN 37914

PIONEER ELECTRONICS
5000 Airport Plaza Dr., Box 1540
Long Beach, CA 90801

QUASAR CO.
9401 W. Grand Ave.
Franklin Park, IL 60131

SANYO ELECTRONICS
1200 W. Artesia Blvd.
Compton, CA 90220

TELETONE CO. INC.
444 S. 9th Ave.
Mt. Vernon, NY 10550

TOSHIBA AMERICA
92 Totawa Rd.
Wayne, NJ 07470

UMC CORP.
3843 Caisen St.
Torrance, CA 90503

YORX ELECTRONICS CORP
405 Minnisink Rd.
Totowa, NJ 07512

PORTABLE TAPE RECORDERS

AIWA AMERICA INC.
800 Corporate Dr.
Mahwah, NJ 07430

AKAI AMERICA LTD.
800 W. Artesia Blvd.
Compton, CA 90220

ALARM INC.
Box 550
Troy, MI 48099

ANOW TRADING CO.
1115 Broadway
New York, NY 10010

DAEWOO ELECTRONCIS
100 Daewoo Pl.
Carlstadt, NJ 07041

DEJAY CORP.
5 Mear Rd.
Holbrook, MA 02343

EMERSON RADIO CORP.
One Emerson Ln.
N. Bergen, NJ 07047

FORTUNE STAR PRODUCTS
12 W. 23rd St.
New York, NY 10010

GENERAL ELECTRIC
Electronics Park, Blvd. 5
Syracuse, NY 13221

GENERAL ELECTRIC
Portsmouth, VA 23705

GOLDSTAR ELECTRONICS
1050 Wall St.
W. Lyndhurst, NJ 07071

GRUNDIG/GR ELECTRONICS
Glenspoint Center E.
Teaneck, NJ 07666

HANABASLHIGA LTD.
39 W. 28th St.
New York, NY 10001

HANIMEX USA INC.
3125 Commercial Ave.
Northbrook, IL 60062

HITACHI SALES CORP.
401 W. Artesia Blvd.
Compton, CA 90220

JORDECKE
1201 Broadway
New York, NY 10001

JULIETTE ELECTRONICS
4565 NW 77th Ave.
Miami, FL 33166

JVC CORP. OF AMERICA
41 Slater St.
Elmwood Park, NJ 07407

K & K MERCHANDISING CORP.
10-27 45th Ave.
Long Island, NY 11101

KENWOOD ELECTRONICS
1315 E. Watson Center Rd.
Carson, CA 90745

KOSS CORP.
4129 N. Port Washington Ave.
Milwaukee, WI 53212

LASONIC ELECTRONICS
1827 W. Valley Blvd.
Alhambra, CA 91813

LLOYDS ELECTRONICS
180 Rariton Center Parkway
Edison, NJ 08818

MAGNAVOX ELECTRONICS
Box 6950
Knoxville, TN 37914

MARANTZ CO.
20525 Nordhoff St.
Chatsworth, CA 91311

MCI
23 NW 8th Ave.
Hollandale, FL 33009

OLYMPUS CORP.
Crossways Park
Woodbury, NY 11797

PANASONIC ELECTRONICS
One Panasonic Way
Secaucus, NJ 07094

PIERRE CARDAN ELECTRONICS
1115 Broadway
New York, NY 10010

PIONEER ELECTRONICS
500 Airport Plaza Dr., Box 1540
Long Beach, CA 90801

QUASAR CO.
9401 W. Grand Ave.
Franklin Park, IL 60131

SAMSUNG CORP.
1250 Valley Brook Ave.
Lyndhurst, NJ 07071

SANSUI ELECTRONICS
301 Mayhill St.
Saddlebrook, NJ 07662

SANYO CORP. OF AMERICA
1050 Arthur Ave.
Elk Grove Village, IL 60007

SANYO ELECTRONICS
1200 W. Artesia Blvd.
Compton, CA 90220

SHARP ELECTRONICS
10 Sharp Plaza
Paramus, NJ 07652

SONY CORP. OF AMERICA
Sony Drive
Park Ridge, NJ 07024

SOUNDESIGN
34 Exchange Pl.
Jersey City, NJ 07302

SYMPHONIC ELECTRONICS
1825 Acacia Blvd.
Compton, CA 90220

TATUNG CO. OF AMERICA
2850 El Presidio St.
Long Beach, CA 90810

TOSHIBA CORP.
82 Totowa Rd.
Wayne, NJ 07470

TZL International Corp.
1523 NW 79th Ave.
Miami, FL 33126

VCM Corp.
3848 Carsen St.
Torrance, CA 90503

Wald Sound
11131 Dora St.
Sun Valley, CA 91352

Windsor Industries Inc.
131 Executive Blvd.
Farmingdale, NY 11735

Yamaha Electronics Corp.
6660 Orangethorpe Ave.
Buena Park, CA 90620

Yorx Electronics Corp.
405 Minnisink Rd.
Totowa, NJ 07512

TAPE DECKS

Adcom
11 Elkins Rd.
E. Brunswick, NJ 08816

Aiwa America Inc.
800 Corporate Dr.
Mahwah, NJ 07430

Akai America Ltd.
800 W. Artesia Blvd.
Compton, CA 90220

Alarm, Inc.
Box 550
Troy, MI 48099

Analog Digital Systems
One Progress Way
Wilmington, MA 01887

Bang & Olufsen of America
1150 Feehanville Dr.
Mt. Prospect, IL 60056

Fortune Star Products
1200 23rd St.
New York, NY 10010

General Electric
Portsmouth, VA 23705

Harman Kardon
240 Crossways Park
W. Woodbury, NY 11797

Hitachi Sales Corp.
401 W. Artesia Blvd.
Compton, CA 90220

JVC Co. of America
41 Slater Dr.
Elmwood Park, NJ 07407

Kenwood Electronics
1315 E. Watson Center Rd.
Carson, CA 90745

Kyocera International
7 Powder Horn Dr.
Warren, NJ 07060

Luxman Division
3102 Kashiwa St.
Torrance, CA 90505

Marantz Co.
20525 Nordhoff St.
Chatsworth, CA 91311

Mitsubishi Electronics Sales
3030 E. Victoria
Rancho Dominguez, CA 90221

NAD Inc.
675 Canton St.
Norwood, MA 02062

NAKAMICHI USA CORP.
19701 S. Vermont Ave.
Torrance, CA 90502

NIKKO AUDIO
5830 S. Triangle Dr.
Commerce, CA 90040

ONKYO USA CORP.
200 Williams Dr.
Ramsey, NJ 07446

PANASONIC ELECTRONICS GROUP
One Panasonic Way
Secaucus, NJ 07094

PIONEER ELECTRONICS
5000 Airport Plaza Dr.
Long Beach, CA 90801

RCA CONSUMER ELECTRONICS
800 N. Sherman Dr.
Indianapolis, IN 46201

SAE
1734 Gage Rd.
Montebello, CA 90640

SANSUI ELECTRONICS CORP.
1250 Valley Brook Ave.
Lyndhurst, NJ 07071

SANSUNG ELECTRONICS
301 Mayhill St.
Saddlebrook, NJ 07662

SANYO ELECTRONICS
1200 W. Artesia Blvd.
Compton, CA 90220

H. H. SCOTT CO.
20 Commercial Way
Woburn, MA 01888

SHARP ELECTRONICS
10 Sharp Plaza
Paramus, NJ 07652

SONY CORP.
Sony Dr.
Park Ridge, NJ 07074

STUDER REVOX AMERICA
1245 Elm Hill Pike
Nashville, TN 37210

TANDBERG OF AMERICA
Labriola Ct.
Armonk, NY 10504

TEAC CORP. OF AMERICA
7733 Telegraph Rd.
Montebello, CA 90640

TECHNICS
One Panasonic Way
Secaucus, NJ 07094

TOSHIBA AMERICA INC.
82 Totawa Rd.
Wayne, NJ 07470

TZL INTERNATIONAL
1523 NW 79th Ave.
Miami, FL 33126

VECTOR RESEARCH
20600 Nordhoff St.
Chatsworth, CA 91311

YAMAHA ELECTRONICS
6660 Orangethorpe Ave.
Buena Park, CA 90620

VCRs

AKAI AMERICA LTD.
800 W. Artesia Blvd.
Compton, CA 90220

CANNON USA INC.
One Cannon Plaza
Lake Success, NY 11042

DENNON AMERICA INC.
222 New Rd.
Parsippany, NJ 07504

FISHER CORP.
21314 Lassen St.
Chatsworth, CA 91311

GENERAL ELECTRIC CONSUMER ELEC.
Portsmouth, VA 23705

GOLDSTAR ELECTRONICS INC.
1050 Wall St.
W. Lyndhurst, NJ 07071

HITACHI SALES CORP.
401 W. Artesia Blvd.
Compton, CA 90220

JENSEN INC.
4136 W. United Parkway
Scheller Park, IL 60176

JVC CO. OF AMERICA
41 Slater Dr.
Elmwood Park, NJ 07401

KENWOOD ELECTRONICS
1315 E. Watson Rd.
Carson, CA 90745

LLOYDS ELECTRONICS
180 Rariton Center Pkwy.
Edison, NJ 08818

MAGNAVOX CONSUMER ELECTRONICS
Box 6950
Nashville, TN 37914

MITSUBISHI CORP.
3030 E. Victoria
Rancho Dominguez, CA 90221

NEC HOME ELECTRONICS
1401 W. Ester Ave.
Elk Grove Village, IL 60007

PANASONIC CONSUMER ELECTRONICS
1200 W. Artesia Blvd.
Secaucus, NJ 07094

PHILCO CONSUMER ELECTRONICS
Box 6950
Knoxville, TN 37914

QUASAR CO.
9401 W. Grand Ave.
Franklin Park, IL 60131

RCA CONSUMER ELECTRONICS
600 N. Sherman Dr.
Indianapolis, IN 46201

SAMSUNG ELECTRONICS INC.
301 Mayhill St.
Saddlebrook, NJ 07662

SANSUI ELECTRONICS CORP.
1250 Valley Brook Ave.
Lyndhurst, NJ 07071

SANYO CORP.
1050 Arthur Ave.
Elk Grove Village, IL 60007

SANYO ELECTRONIC INC.
One Panasonic Way
Compton, CA 90220

SHARP ELECTRONICS
10 Sharp Plaza
Paramus, NJ 07652

SONY CORP. OF AMERICA
Sony Dr.
Park Ridge, NJ 07074

SYLVANIA CONSUMER ELECTRONICS
Box 6950
Knoxville, TN 37914

SYMPHONIC ELECT. CORP.
1825 Acacia
Compton, CA 90220

TATUNG CO. OF AMERICA
2850 El Presidio St.
Long Beach, CA 90810

TECHNICS ELECTRONIC CORP.
353 Rt. 46 W.
Fairfield, NJ 07470

TOSHIBA AMERICA INT.
82 Totowa Rd.
Wayne, NJ 07470

TOYOMENKA INC.
357 County Ave.
Secaucus, NJ 07094

VECTOR RESEARCH
20600 Nordhoff St.
Chatsworth, CA 91311

ZENITH ELECTRONICS CORP.
1000 Milwaukee Ave.
Glenview, IL 60025

Glossary

acoustic suspension Air-suspension (AS) speakers are sealed in an enclosure or box to produce natural, low-distortion base output. Greater driving power is needed with these less-efficient speaker systems.

air suspension Another name for an acoustic-suspension speaker.

amp Abbreviation for amplifier.

ANRS A noise-reduction system that operates on principles that are similar to the Dolby system found in JVC products.

APC Automatic power control is the circuit that keeps the laser-diode optical output at a constant level in the CD player.

audio/video control center The central control system that controls all audio and VCR operations.

auto eject The tape player feature that automatically ejects the cassette at the end of the playing time.

auto focus AF is the focus servo that moves the objective lens up or down to correct the focus of the CD player.

auto record level Automatic control of the recording level.

auto reverse The ability of the cassette player to automatically reverse directions to play other side of the tape.

auto tape selector Automatic bias and equalization when the cassette is inserted into the tape deck.

azimuth The angle at which the tape head meets the moving tape. A loss of high-frequency response is often caused by improper azimuth adjustment.

azimuth control A control to adjust the angle of the tape control to correct misalignment in the auto stereo tape player.

baffle The board on which the speakers are mounted.

balance The control in the stereo amp that equalizes the output audio in each channel.

bass reflex A bass-reflex system vents backward sound waves through a tuned vent or port to improve bass response.

bias A high-frequency current applied to the tape-head winding to prevent low distortion and noise while recording.

block diagram A diagram that shows the different stages of a system.

booster amplifier A separate amplifier that is connected between the main unit and the speakers in a car stereo system.

bridging Combining both stereo channels of the amp to produce a mono signal with almost twice the normal power rating in a car stereo system.

cabinet A box that contains speakers or electronic equipment.

capstan The shaft that rotates against the tape at a constant rate of speed and moves the tape past the tape heads. In the cassette player, a rubber pinch roller holds the tape against the capstan.

cassette radio The combination of an AM/FM tuner, amplifier, and cassette player in one unit.

cassette tuner A tuner and cassette deck in one chassis.

CH The abbreviation for channel. The stereo component has two channels (left and right). '

channel separation The degree of isolation between the left and right channels, often impressed in decibels. The higher the decibel values, the better the separation.

chassis The framework that holds the working parts in the amplifier, tuner, radio, cassette, CD player, or VCR recorder. The chassis could be metal, plastic, or a PC board.

chips Chip devices contain resistors, multi-layer ceramic chip capacitors, mini-mould chip transistors, mini-mould chip diodes, and mini-mould chip ICs.

clipping Removing or cutting off the signal from a waveform that contains distortion, which can be seen on the oscilloscope. Excessive power results in distortion.

coaxial speaker A speaker with two drivers mounted on the same frame. The tweeter is mounted in front of the woofer speaker. Usually, coaxial speakers are used in the car audio system.

compact disc The compact-disc (CD) player plays a small disc of digitally encoded music. The CD provides noiseless high-fidelity music on one side of a rainbow-like surface.

CPU A computer-type processor used in the master and control mechanism circuits of a CD player.

crossover A filter that divides the signal to the speaker into two or more frequency ranges. The high frequencies go to the tweeter and the low frequencies go to the woofer.

crosstalk Leakage of one channel into the other. Improper adjustment of the tape head might cause crosstalk between two different tracks.

D/A converter In the CD player, the device that converts the digital signal to an analog or audio signal.

dc Direct current is found in automobile battery systems, and also after the ac has been filtered and rectified in low-voltage power supplies.

decibel   The decibel (dB) is a measure of gain, the ratio of the output power or voltage, with respect to the input (expressed in log-units).

de-emphasis   A form of equalization in FM tuners to improve the overall signal-to-noise ratio while maintaining the uniform frequency response. The de-emphasis stage follows the D/A converter in a CD player.

dew   A warning light that might come on in a VCR or camcorder. It indicates too much moisture at the tape head.

digital   Within tuners, the digital system is a very precise way to lock in a station without drifting. Digital recording is found in the compact discs.

direct drive   A direct-drive motor shaft is connected to a spindle or capstan/fly-wheel. The CD rests directly on the disc or spindle motor in CD players.

disc holder   The disc holder or turntable sits directly on top of the motor shaft in the CD player.

dispersion   1. The spread of speaker high frequencies, measured in degrees. 2. The angle by which the speaker radiates its sound.

distortion   In a simple sine-wave signal, distortion appears as multiples (harmonics of the input frequency). A type of distortion is the clipping of the audio signal in the audio amplifier.

DNR   Dynamic noise reduction is a noise-reduction system that reduces the high frequencies when the signal is at a low level.

Dolby noise reduction   A type of noise reduction that works by increasing the treble sounds during recording and decreasing them during playback, thus restoring the signal to the original level and eliminating tape hiss.

driver   1. In a speaker system, each separate speaker is sometimes called a driver. 2. The loading, feed, and disc motors might be driven by transistor or IC drivers.

drive system   The motors, belts, and gears that drive the capstan/flywheel in cassette tape or CD players.

dropout   In tape machines, dropouts occur when the tape does not contact the tape head for an instant. Dropouts occur in the compact disc because of dust, dirt, or deep scratches on the plastic disc.

dual capstan   Dual capstans and flywheels are used in auto-reverse cassette players and can play tapes in both directions.

dynamic   A dynamic speaker has a voice coil that carries the signal current with a fixed magnetic field (PM magnet), and moves the coil and cone. The same principle applies to the human ear or to headphones.

dynamic range   The ratio between the maximum signal-level range and the minimum level, expressed in decibels (dB).

E/F balance   After changing the optical laser pick-up assembly, the balance of E/F diodes must be adjusted for tracking-error detection in the CD player.

EFM signal   The EFM signal is fed into signal processor LS1 of the CD player.

electronic speed control   An electronic method of controlling the speed of the motor.

electrostatic   An electrostatic speaker, headphone, or microphone, which uses a thin diaphragm with a voltage applied to it. The electrostatic field is varied by the voltage, which moves the diaphragm to create sound.

equalizer   A device to change the volume of certain frequencies, in relation to the rest of the frequency range. Sliding controls can be found in auto-radio and cassette-player equalizers.

erase head   A magnetic component with applied voltage or current to remove the previous secondary or noises on the tape. The erase head is mounted ahead of the regular R/P head.

extended play   EP refers to the six hours of playing time that is obtainable with a T-120 VHS cassette played in a VCR machine.

eye pattern   The RF signal waveform at the RF amplifier in a CD player. The waveform is adjusted to a clear and distinct diamond-shaped pattern.

fader   A control in auto radio or cassette players to control the volume balance between front and rear speakers.

fast forward   The motor in the cassette, VCR, or CD player can rotate in faster with a higher voltage applied to the motor terminals or when larger idler pulleys pushed into operation.

filter   A circuit that selectively attenuates certain frequencies, but not others. The large electrolytic capacitor in the low-voltage power supply is sometimes called a *filter capacitor*.

flutter   A change in the speed of a tape transport, also known as *wow*.

focus error   The output from the four optosensing elements are supplied to the error-signal amplifier and a zero output is produced. The error amp corrects the signal voltage and sends to the servo IC to correct the focus in the CD player.

folded horn speaker   The system that efficiently forces the sound of the driver to take a different path to the listener.

frequency response   The range of frequencies that a given piece of equipment can pass to the listener. The frequency response of a given amplifier might be 20 Hz to 20 kHz.

gain   The amplification of an electronic signal. Gain is given in decibels.

gain control   A control to adjust the amount or boost the amount of signal.

gap   The critical distance between the pole pieces of the tape head. The gap area might be full of oxide, which would cause weak, distorted, or noisy reception.

glitch   A form of audio or video noise or distortion that suddenly appears and disappears during VCR operation.

graphic equalizer   An equalizer with a series of sliders that provides a visual graphic display.

ground   A point of zero voltage within the circuit. The common ground might be a metal chassis in the amplifier. American-made cars have a negative-ground polarity.

harmonic   The addition of harmonics not present in the original recording. A good tape player should have less than 1% distortion.

harmonics   A series of multiples of the fundamental frequency. Harmonics help determine the total quality of a sound.

head A magnetized component with a gap area that picks up signals from the revolving tape.

hertz Hertz (Hz) are the number of cycles per second (CPS), the unit of frequency.

hiss The annoying high-frequency background noise in tapes and record players.

hum A type of noise that originates from power lines, caused mainly by poor filtering in the low-voltage power supply. Hum and vibrating noise might be heard in transformers or motors that have loose particles or laminations.

idler A wheel found in tape players to determine the speed of the capstan/flywheel or turntables in the cassette player.

impedance The degree of resistance (in ohms), that an electrical current will encounter in a given circuit or component. A speaker might have an impedance of 2, 4, 8, 16, or 32 Ω.

infinite baffle A completely sealed box that encloses the speakers.

integrated amp A single component that combines the circuitry and functions of the preamp and power-output amplifier.

integrated circuit An IC is a single component that has many parts. ICs are found throughout most cassette players, amplifiers, VCRs, and CD players.

interlock A safety interlock device used in the CD player to load the disc.

ips Inches per second, the measurement of cassette-tape speed.

jack The female part of a plug and receptacle.

kilohertz 1 kHz is equal to 1,000 Hz.

laser assembly The assembly that contains the laser diodes, focus, and tracking coils in a CD player.

laser current Low laser current might indicate a laser-diode assembly in the CD player is defective.

laser diodes The diodes that pick up the coded information from the disc along with the optical pick-up assembly in a CD player.

LED Light-emitting diodes are used for optical readouts and displays in electronic equipment.

level 1. The strength of a signal. 2. The alignment of the tape head with the tape.

line Line output or input jacks are used in the amplifier, cassette, or CD player. The line signal is usually a high-level signal.

loading motor The motor in CD, VCR, and camcorders that moves the tray or lid out and in so that the disc or cassette can be loaded.

long play LP is a speed on the VCR that provides four hours of recording on a 120-minute VHS cassette.

loudness The volume of sound. Loudness is controlled by a volume control.

LSI Large-scale integrated circuits include processors, ICs, and CPUs that are found in VCRs, camcorders, and disc players.

magnetic Metal attraction. The magnetic coil might be found in the VOM or VTVM.

megahertz 1 MHz is equal to 1,000 kHz.

memory The program memory of a CD player.

metal tape The high-frequency response and maximum-output level are greatly improved with metal tape. Pure metal cassettes are more expensive than the regular oxide cassettes.

microprocessor   A multifunction chip found in most of today's electronic products. They are used in tape decks, transports, memory operations, and cassette, CD, and VCR players.

monitor   To compare signals. A stereo amplifier can be monitored to compare the signal with the defective channel.

monophonic   One channel of audio, such as in a single speaker.

MOSFET   Metal oxide semiconductor field-effect transistor.

multiplex   A multiplex (MPX) demodulator in the FM tuner or receiver converts a single-carrier signal into two stereo channels of audio.

mute switch   The mute switch might be a transistor in the audio-output line circuit of a CD or cassette player.

noise   Any unwanted signal that is related to the desired signal. Noise can be generated during the record and play functions in a cassette player. A defective transistor or IC could cause a frying noise in the audio.

NR   Noise reduction.

optical lens   The lens located in the pick-up head of a CD player. Clean the lens with solution and a photograph dry-cleaning brush.

output power   The output power of an amplifier, rated in watts.

oxide   The magnetic coating compound of the recording tape or cassette. The excess oxide should be cleaned off of the tape heads, pinch rollers, and capstans for good music reproduction.

passive radiator   A second woofer cone that is added without a voice coil in the speaker cabinet. The pressure created by the second cone produces heavy bass tones.

pause control   A feature to stop the tape movement without switching the machine. The pause control is used in cassette, VCR, and CD players.

PBX   The noise-reduction system in which the program is compressed before being recorded and expanded in playback.

peak   The level of power or signal. A peak indicator light shows that the signal levels are exceeding the recorder's ability to handle the peaks without distorting.

phase   Sound waves are in sync with one another. Speakers should be wired in phase.

pick-up motor   The pick-up, SLED, or feed motor is used to move the pick-up assembly in the radial direction or toward the outer edge of the disc.

pitch control   A control that changes the speed of the control motor.

PLL   The phase-locked loop (PLL) VCO circuit is used in the digital-control processor of the CD player with a crystal.

port   An opening in a speaker enclosure or cabinet. The port permits the back bass radiation to be combined with the front radiation for total response.

power   The output power of any amp is given in watts. A low-voltage power supply provides voltage to other circuits.

preamplifier   The amp within the cassette player that takes the weak signal from the tape head and amplifies it for the AF stages.

rated power bandwidth The frequency range over which the amplifier supplies a certain minimum power factor, usually from 20 to 20,000 Hz.

recording-level meter The meter that indicates how much signal is being recorded on the tape with vane type, LED, or a fluorescent panel.

reject lever A lever that rejects or deletes a given track in a cassette or a record on the record changer.

remote control A means to operate the receiver, CD player, cassette/tuner, or VCR from a distance. Today, most remote-controlled transmitters are infrared type.

repeat button The button that replays the same track of music on the CD player.

RF A radio-frequency signal.

ribbon speaker A high-frequency driver or tweeter speaker that uses a ribbon material suspended in a magnetic field to generate sound current when current is passed through it.

sample hold The circuits that are found in each stereo channel after the digital-to-analog (D/A) processor within the CD player.

saturation Recording tape is saturated when it cannot hold anymore magnetic information.

self erase A degrading or partial erasure of information on magnetic tape.

self-powered speakers A speaker with a built-in amplifier.

separation The separation of two stereo channels. Placement of the stereo speakers can provide good or poor stereo separation.

servo The tracking circuits that keep the laser pickup in the grooves at all times.

servo control The servo control IC that controls the focus and tracking coils in CD players.

signal processing In the CD player, converting the processing laser signals to audio with preamps and signal processors.

signal-to-noise ratio The ratio (S/N) of the loudest signal to noise. The higher the signal-to-noise ratio, the better the sound.

skewing A form of visual distortion or bend at the upper part of the picture of the VCR player.

SMD Surface-mounted devices are small surface-mounted components: resistors, capacitors, transistors, and ICs that are mounted flat on the PC wiring side.

solenoid A switch that consists of an electric coil with an iron-core plunger that is pulled inside the coil by the magnetic field. Solenoids are usually found in auto radios, cassette, tape, and CD players.

speaker enclosure The cabinet in which speakers are mounted.

spindle motor The disc or turntable motor revolves.

standard play SP is the speed at which a two-hour (T-120) VHS cassette plays on VCR machine.

subwoofer A speaker that is designed to handle very low frequencies below 150 Hz.

test cassette The recorded signals on a test cassette that are used for alignment and adjustment procedures on the cassette player.

test disc A CD that is used to make alignments and adjustments in the CD players.

tone control   A circuit that is designed to increase or decrease the amplification in a specific frequency range.

tracking servo   The IC processor that keeps the laser beam in focus and tracking correctly.

tray   The loading tray in which the CD to be played is placed.

tweeter   A high-frequency driver speaker.

VCR   Video cassette recorder.

vented speaker system   Any speaker cabinet with a hole or port to let the back waves of the woofer speaker escape. A *bass reflex* is a type of vented speaker system.

VHS   The system used today in most VCRs.

voice coil   The coil of wire that is wound over the end of the cone of the speaker in which the amplifier output is connected. The electrical signal is converted to mechanical energy to create audible sound waves.

watts   The practical unit of electric and other power.

woofer   The largest speaker in a speaker system. The one that reproduces the low frequencies.

wow   A slow-speed fluctuation in tape speed. Fast-speed variation is called *flutter*.

Index

W